A BRIEF WELCOME TO THE UNIVERSE

A BRIEF WELCOME TO THE UNIVERSE

A Pocket-Sized Tour

NEIL DEGRASSE TYSON,
MICHAEL A. STRAUSS,
AND J. RICHARD GOTT

PRINCETON UNIVERSITY PRESS
PRINCETON AND OXFORD

Published by Princeton University Press
41 William Street, Princeton, New Jersey 08540
6 Oxford Street, Woodstock, Oxfordshire OX20 1TR

press.princeton.edu

Library of Congress Cataloging-in-Publication Data

Names: Tyson, Neil deGrasse, author. | Strauss, Michael Abram, author. | Gott, J. Richard, author.
Title: A brief welcome to the universe : a pocket-sized tour / Neil deGrasse Tyson, Michael A. Strauss, and J. Richard Gott.
Description: Princeton, New Jersey : Princeton University Press, [2021] | Includes index.
Identifiers: LCCN 2021018408 (print) | LCCN 2021018409 (ebook) | ISBN 9780691219943 (paperback) | ISBN 9780691223612 (ebook)
Subjects: LCSH: Cosmology—Popular works. | Astrophysics—Popular works. | BISAC: SCIENCE / Physics / Astrophysics | SCIENCE / Physics / General
Classification: LCC QB982 .T963 2021 (print) | LCC QB982 (ebook) | DDC 523.1—dc23
LC record available at https://lccn.loc.gov/2021018408
LC ebook record available at https://lccn.loc.gov/2021018409

British Library Cataloging-in-Publication Data is available

Editorial: Ingrid Gnerlich, María García, Whitney Rauenhorst
Production Editorial: Mark Bellis
Cover Design: Karl Spurzem
Production: Jacqueline Poirier
Publicity: Sara Henning-Stout and Kate Farquhar-Thomson
Copyeditor: Kathleen Kageff

This book has been composed in Adobe Text Pro and Futura

Printed on acid-free paper. ∞

Printed in the United States of America

10 9 8 7 6 5 4 3 2 1

CONTENTS

A NOTE TO THE READER

A Brief Welcome to the Universe: A Pocket-Sized Tour is the spirit essence of our larger collaboration *Welcome to the Universe: An Astrophysical Tour*. If that book was an all-you-can-eat cosmic banquet, this book offers appetizer portions, intended to stimulate your appetite for more.

A BRIEF WELCOME TO THE UNIVERSE

CHAPTER 1

SIZE AND SCALE OF THE UNIVERSE

Neil deGrasse Tyson

We begin with the solar system. Ascend to the stars. Then reach for the galaxy, the universe, and beyond.

The universe. It's bigger than you think. It's hotter than you think. It is denser than you think. It's more rarified than you think. Everything you think about the universe is less exotic than it actually is. Let's get some numerical machinery together before we begin. Start with the number 1. You've seen this number before. There are no zeros in it. If we wrote this in exponential notation, it is ten to the zero power, 10^0. The number 1 has no zeros to the right of that 1, as indicated by the zero exponent. Moving onward, the number

10 can be written as 10 to the first power, 10^1. Let's go to a thousand—10^3. What's the metric prefix for a thousand? *Kilo-* kilogram—a thousand grams; kilometer—a thousand meters. Let's go up another three zeros, to a million, 10^6, whose prefix is *mega-*. Maybe this is the highest they had learned how to count at the time they invented the megaphone; perhaps if they had known about a billion, by appending three more zeroes, giving 10^9, they would have called them "gigaphones."

Do you know how big a billion is? What kinds of things come in billions?

Currently we are approaching 8 billion people in the world.

How about Jeff Bezos, the founder of Amazon .com? What's his wealth up to? More than 100 billion dollars. Where have you seen 100 billion? Well, McDonald's: "Over 99 Billion Served." That's the biggest number you ever see in the street. McDonald's never displayed 100 billion, because they allocated only two numerical slots for their burger count, and so, they just stopped at 99 billion. After that, they pulled a Carl Sagan on us and now say, "billions and billions served."

Take 100 billion hamburgers, and lay them end to end. Start at New York City, and go west. Will you get to Chicago? Of course. Will you get to California? Yes. Find some way to float them. This calculation uses the diameter of the bun (4 inches), so it's all about the bun. Now float them across the ocean, along a great circle route, and you will cross the Pacific, pass Australia, the Indian Ocean, Africa, and across the Atlantic Ocean, finally arriving back in New York City. That's a lot of hamburgers. But you have some left over after you have circled Earth's circumference. So, you make the trip all over again, 215 more times. Afterward, you still have some left. You're bored circumnavigating Earth, so you stack what remains. How high do you go? You'll go to the Moon, and back, with stacked hamburgers (each 2 inches tall) after you've already been around the world 216 times. Only then will you have used your 100 billion hamburgers. That's why cows are scared of McDonald's. By comparison, the Milky Way galaxy has about 300 billion stars. Perhaps McDonald's is gearing up for the cosmos.

When you are 31 years, 7 months, 9 hours, 4 minutes, and 20 seconds old, you've lived your billionth second. I'm just geeky enough to have celebrated that moment in my life with a fast sip of champagne.

Let's keep going. What's the next step up? A trillion: 10^{12}. We have a metric prefix for that: *tera-*. You can't count to a trillion. If you counted one number every second, it would take you 1,000 times 31 years—31,000 years, which is why we don't recommend doing this, even at home. A trillion seconds ago, cave dwellers—troglodytes—were drawing pictures on their living-room walls.

At New York City's Rose Center of Earth and Space, a spiral ramp timeline of the universe begins at the Big Bang and displays 13.8 billion years. Uncurled, it's the length of a football field. Every step you take spans 50 million years. You get to the end of the ramp, and you ask, where are we? Where is the history of our human species? The entire period of time, from a trillion seconds ago to today, from graffiti-prone cave dwellers until now, occupies only the thickness of a single strand of human hair, which we have mounted at the end of that timeline. You think we live long lives; you think civilizations last a long time? No. Not relative to the cosmos itself.

What's next? 10^{15}. That's a quadrillion, with the metric prefix *peta-*. Between 1 and 10 quadrillion ants live on (and in) Earth, according to Harvard biologist E. O. Wilson.

Then comes 10^{18}, a quintillion, with metric prefix *exa-*. That's the estimated number of grains of sand on ten large beaches.

Up another factor of 1,000 and we arrive at 10^{21}, a sextillion. We have ascended from kilometers to megaphones to McDonald's hamburgers to Cro-Magnon artists to ants to grains of sand on beaches, until finally arriving here: more than 10 sextillion—

the number of stars in the observable universe.

There are people, who walk around every day, asserting that we are alone in this cosmos. They simply have no concept of large numbers, no concept of the size of the cosmos. Later, we'll learn more about what we mean by the *observable universe*, the part of the universe we can see.

While we're at it, how about a number much larger than 1 sextillion—10^{81}? It's the number of atoms in the observable universe. Why would you ever need a number bigger than that? What "on Earth" could you be counting? How about 10^{100}, a nice round-looking number. This is called a *googol.* Not to be confused with Google, the internet company that misspelled "googol" on purpose.

There are not enough objects in the universe for a googol to count. It is just a fun number. We can write it as 10^{100}, or as is true for all out big numbers, if you don't have superscripts, this works too: 10^100. But you can still use such big numbers for some situations: don't count *things*; instead count the ways things can happen. For example, how many possible chess games can be played? A game can be declared a draw by either player after a triple repetition of a position, or when each has made 50 moves in a row without a pawn move or a capture, or when there are not enough pieces left to produce a checkmate. If we say that one of the two players must declare a draw whenever one of these three things happen, then we can calculate the number of all possible chess games. Rich Gott did this (because that's just the kind of thing he does) and found the answer was a number less than 10^(10^4.4). That's a lot bigger than a googol, which is 10^(10^2). Again, you're not counting things; you are counting possible ways of doing things. In that way, numbers can get very large.

Here's a still bigger number. If a googol is 1 followed by 100 zeros, then how about 10 to the googol power? That has a name too: a *googolplex*. It is 1, with a googol zeroes after it. Can you even write out this number? Nope. You would need a googol zeroes, and a googol is larger than the number of atoms in the universe, then you're stuck writing it this way: 10^{googol}, or $10^{10^{\wedge}100}$ or 10^(10^100).

We're not just wasting your time. Here's a number bigger than a googolplex. Jacob

Bekenstein invented a formula allowing us to estimate the maximum number of different quantum states that could have a total mass and size comparable to our observable universe. Given the quantum fuzziness we observe, that would be the maximum number of distinct observable universes like ours. It's 10^(10^124), which has 10^{24} times as many zeros as a googolplex. These 10^(10^124) universes range from ones that are scary, filled with mostly black holes, to ones that are exactly like ours but where your nostril is missing one oxygen molecule and some space alien's nostril has one more.

A mathematical theorem once contained the badass number 10^(10^(10^34)). It's called *Skewe's number*. And it dwarfs them all.

Time to get a sense of the extremes in the universe.

How about density? You intuitively know what density is, but let's think about density in the cosmos. First, explore the air around us. You're breathing 2.5×10^{19} molecules per cubic centimeter—78% nitrogen and 21% oxygen (plus 1% "other"). When we talk about density here, we're referencing the number of molecules, atoms, or loose particles that compose the material in question.

A density of 2.5×10^{19} molecules per cubic centimeter is likely higher than you thought. What about our best laboratory vacuums? We do pretty well today, bringing the density down to about 100 molecules per cubic centimeter. How about interplanetary space? The solar wind at Earth's distance from the Sun has about 10 protons per cubic centimeter. How about interstellar space, between the stars? Its density fluctuates, depending on where you're hanging out, but regions in which the density falls to 1 atom per cubic centimeter are not uncommon. In intergalactic space, that number is much less: 1 per cubic meter.

We can't get vacuums that empty in our best laboratories. There is an old saying, "Nature abhors a vacuum." People who said that never left Earth's surface. In fact, Nature just *loves* a vacuum, because that's what most of the universe is. When they said "Nature," they were just referring to the base of this blanket of air we call our atmosphere, which does indeed rush in to fill empty spaces whenever it can.

Smash a piece of chalk into smithereens against a blackboard and pick up a fragment. Let's say a smithereen is about 1 millimeter across. Imagine that's a proton. Do you know what the simplest atom is? Hydrogen. Its nucleus contains one proton, and normal hydrogen has an electron occupying a spherically shaped volume that surrounds the proton. We call these volumes orbitals. If the chalk smithereen is the proton, then how big would the full hydrogen atom be? One hundred meters across—about the size of a football field. So atoms are quite empty, though small: about 10^{-10} meters in diameter. That's one ten-billionth of a meter. Only when you get down to 10^{-14} or 10^{-15} meters are you measuring the size of the nucleus. Let's go smaller. We do not yet know the diameter of the electron. It's smaller than we are able to measure. However, superstring theory suggests that it may be a tiny vibrating string as small as 1.6×10^{-35} meters in length. So matter is an excellent repository of empty space.

Now let's go the other way, climbing to higher and higher densities. How about the Sun? It's quite dense (and crazy hot) in the center, but much less dense at its edge. The average density of the Sun is about 1.4 times that of water. And we know the density of water—1 gram per cubic centimeter. In its center, the Sun's density is 160 grams per cubic centimeter. Yet the Sun is undistinguished in these matters. Stars can (mis) behave in amazing ways. Some expand to get big and bulbous with very low density, while others collapse to become small and dense. In fact, consider the proton smithereen and the lonely, empty space that surrounds it. There are processes in the universe that collapse matter down, crushing it until there's no empty volume between the nucleus and the electrons. In this state of existence, the matter reaches the density of an atomic nucleus. Within such stars, each nucleus rubs cheek to cheek with neighboring nuclei.

The objects out there with these extraordinary properties happen to be made mostly of neutrons—a super-high-density realm of the universe.

In our profession, we tend to name things exactly as we see them. Big red stars we call *red giants*. Small white stars we call *white dwarfs*. When stars are made of neutrons, we call them *neutron stars*. Stars we observe pulsing, we call them *pulsars*. In biology they come up with big Latin words for things. MDs write prescriptions in a cuneiform that patients can't understand, then hand them to the pharmacist, who understands the cuneiform. In biochemistry, the most popular molecule has ten syllables—deoxyribonucleic acid. Yet the beginning of all space, time, matter, and energy in the cosmos is simply the *Big Bang*. We are a simple people, with a monosyllabic lexicon. The universe is hard enough, so there is no point in making big words to confuse you further.

Want more? In the universe, there are places where the gravity is so strong that light doesn't come out. You fall in, and you can't come out; these are called *black holes*. Once again, with single syllables, we get the whole job done.

How dense is a neutron star? Cram a herd of 100 million elephants into a Chapstick casing.

In other words, if you put 100 million elephants on one side of a seesaw, and a single Chapstick of neutron star material on the other side, they would balance. That's some dense stuff.

How about temperature? Let's talk hot. Start with the surface of the Sun. About 6,000 kelvins—6,000 K (a temperature in kelvins is equal to its temperature in degrees centigrade + 273). That will vaporize anything you give it. That's why the Sun is gas, because that temperature vaporizes everything. By comparison, the average temperature of Earth's surface is a mere 287 K.

How about the temperature at the Sun's center? As you might guess, the Sun's center is hotter than its surface. The Sun's core is about 15 million K.

Let's go cool. What is the temperature of the whole universe? It does indeed have a temperature—left over from the Big Bang. In the beginning, 13.8 billion years ago, all the space, time, matter, and energy you can see, out to 13.8 billion light-years, was crushed together. (A light-year is the distance light, traveling at 300,000 kilometers a second, can travel in a year—about

10 trillion kilometers.) The nascent universe 1 second after its birth was hot, about 10 billion K, a seething cauldron of matter and energy. Cosmic expansion since then has cooled the universe down to a mere 2.7 K.

Today we continue to expand and cool. As unsettling as it may be, all data show that we're on a one-way trip. We were birthed by the Big Bang, and we're going to expand forever. The temperature will continue to drop, eventually becoming 2 K, then 1 K, then half a kelvin, asymptotically approaching absolute zero. Ultimately, its temperature may bottom out at about 7×10^{-31} K (that's 0.7 million-trillion-trillionths of a degree above absolute zero) because of an effect discovered by Stephen Hawking that we will discuss in chapter 8. But that fact brings no comfort. Stars will finish fusing all their thermonuclear fuel, and one by one they will blink out, disappearing from the night sky. Interstellar gas clouds do make new stars, but of course this depletes their gas supply. You start with gas, you make stars, the stars age, and they leave behind a corpse—the dead end-products of stellar evolution: black holes, neutron stars, and white dwarfs. This keeps going until all the lights of the galaxy turn off, one by one. The galaxy goes dark. The universe goes dark. This leaves black holes that emit only a feeble glow of light—again predicted by Stephen Hawking.

And so the cosmos ends. Not in fire, but in ice. And not with a bang, but with a whimper.

Have a nice day! And, welcome to the universe.

CHAPTER 2

PLUTO'S PLACE IN THE SOLAR SYSTEM

Neil deGrasse Tyson

The Rose Center of the American Museum of Natural History in New York City, where I serve as director, contains an 87-foot-diameter sphere in a glass cube. The sphere houses the dome of the Hayden Planetarium in its upper half and a Big Bang theater in its belly. The plan emphasizes that the universe loves spheres, recognizing that the laws of physics conspire to make things round, from stars to planets to atoms. Beginning with an architectural structure that is round, we put it to work as an exhibit element, allowing us to compare the sizes of things in the universe. A walkway girds the 87-foot-diameter sphere where we invite you to envision the "Scales of the Universe." As you start out, imagine the planetarium sphere is the entire observable universe. On the railing sits a model, about the size of your fist, showing the extent of our Virgo supercluster, containing thousands of galaxies, including the Milky Way.

At a subsequent stop, midway along the "Scales of the Universe" walkway, the big sphere represents the Sun with models of the planets placed next to it, in correct size relative to the Sun as the big sphere. This exercise continues, comparing smaller and smaller scales. When the planetarium sphere is the hydrogen atom, we show a dot the size of its nucleus—1/130 of an inch across, revealing and affirming that most of the atom's volume is empty space.

At the place in the walkway where you stand to compare the big sphere, representing the Sun, to the sizes of the planets, we omitted Pluto. That may seem rude, but we had good reasons. That's where all the trouble with Pluto started.

A reporter, visiting a year after the exhibit opened, noticed that Pluto was missing from the display of the relative sizes of planets and decided to make a big deal of it, writing a front-page story about it in the *New York Times*, and that's when all hell broke loose. Here's the fast version what we did and why we did it.

At the beginning of the twentieth century, there were eight known planets: Mercury, Venus, Earth, Mars, Jupiter, Saturn, Uranus, and Neptune. Clyde Tombaugh discovered Pluto in 1930. As late as 1962, Allen's *Astrophysical Quantities* listed Pluto's mass as: 80% that of Earth—with a question mark, indicating it was a best guess. Seems like a planet.

In 1978, a moon of Pluto was discovered—Charon. Charon and Pluto are orbiting their common center of mass, as they should; from observations of the orbit and using *Newton's law of gravity*, we could measure Pluto's mass. The result? Pluto's mass is a mere 1/500 that of Earth, tiny relative to other planets.

By the way, Newton was badass. His law of gravity said two bodies attracted each other with a force that was proportional to the product of their masses divided by the square of the distance between them. It explained the observed elliptical orbits of planets. He derived this formula before he turned 26. He discovered, amazingly, that the colors of the rainbow when combined gave you white light. He invented calculus and the reflecting telescope. He did all this.

So, what do we do about Pluto? Pluto is the smallest planet, by far. There are seven *moons* in the solar system bigger than Pluto, including Earth's moon. Pluto's orbital path is so elliptical it's the only planet whose orbit crosses the orbit of another planet. Pluto is made mostly of ice—55% by volume. We have a word for icy things in the solar system—comets. Pluto shares a lot of features with comets. But it wasn't zooming in close to the Sun and then swinging back out, as most comets do. When an icy comet comes close to the Sun, the comet outgasses vapor and dust, producing a long tail. Comets are small (Halley's comet for example is 7 kilometers in diameter) but can have tails millions of kilometers long. Pluto never gets that close to the Sun, so it doesn't do that. Despite its atypical features, people were happy to keep Pluto within our definition of planet.

In the Rose Center, however, we wanted to future-proof our exhibits as much as possible. Trend lines in planetary exploration mattered to us greatly. Pluto is more different from Mercury, Venus, Earth, and Mars than any of them are from one another. Mercury, Venus, Earth, and Mars are all small and rocky. That's one family.

Mercury, the planet nearest the Sun, has a large iron core, only a trace of an atmosphere, and a cratered surface. Venus is covered in clouds. Beneath the cloud cover it has dramatic mountain ranges, and a few craters. Venus, just a tad smaller than Earth, has a thick atmosphere of carbon dioxide (CO_2), with a scorching greenhouse effect giving an intolerably high surface temperature. Mars is smaller than both Earth and Venus but larger than Mercury. It retains a thin atmosphere of CO_2 that produces only a tiny greenhouse effect. This, coupled with its larger distance from the Sun, makes Mars much colder than Earth. The atmospheric pressure on the surface of Mars is about 1/100 that of Earth. It has dark areas on its surface, made of basalt rock where lava once flowed, not covered by sand and dust. The red areas, making Mars the "red planet," are deserts, deriving their color from the ubiquity of rusty, iron-infused dust and sand. Mars has a rift valley that could span the United States from coast to coast. It has an extinct volcano, Olympus Mons, that is 70,000 feet high—the largest mountain in the solar system. Mars has two polar caps composed mostly of water ice, with a frosting of dry ice (frozen CO_2) on top—in winter on the north polar cap, but year-round on the south polar cap. Mars is the most habitable of the planets other than Earth.

What else is out there? We've got Jupiter, Saturn, Uranus, and Neptune. They are all big and gaseous. That's another family. Once again, they have more in common with one another than any one of them has with Pluto.

Jupiter orbits beyond Mars. It is the largest planet, at 11 times the diameter of Earth. Still, Jupiter is tiny compared with the Sun, which is 109 times the diameter of Earth. Jupiter is composed mostly of hydrogen and helium, quite similar to the Sun. We will see later that these are the two primary birth ingredients of the entire universe.

Jupiter's outer atmosphere contains methane and ammonia clouds. Jupiter has cloud belts, and its Great Red Spot is a storm that has raged for more than 300 years. Saturn is similar to Jupiter but is surrounded by a magnificent set of rings composed of icy particles that orbit the planet. Uranus and Neptune are similar to Jupiter and Saturn, but smaller versions. Neptune has winds reaching 1,500 miles per hour.

The rocky terrestrial planets form in the inner solar system, where, warmed by the Sun, light elements such as hydrogen and helium are heated to high enough temperatures that they can escape the planet's gravity. The gas giant planets, formed in the outer solar system, are colder, retaining all their hydrogen and helium and therefore becoming very massive.

Pluto simply doesn't fit in to any of these family photos. Over the decades, we've just been kind to Pluto, keeping it in the family of planets, even though we knew in our hearts that it didn't belong. A look at textbooks from the late 1970s (when we finally settled on Pluto's size and mass) and the 1980s shows Pluto was beginning to get lumped together with the comets, the asteroids, and other solar system "debris." Those were the first seeds of the unraveling of Pluto's red-blooded planetary status.

Then in 1992, we found another object in the outer solar system—another icy body beyond Neptune. Since then, we've discovered more than two thousand of these objects. What are their orbits like? They're all beyond Neptune, and many have orbital tilts and elliptical shapes that resemble Pluto's orbit. In fact, these newly discovered icy bodies constitute a whole new swath of real estate in our solar system. The astrophysicist Gerard Kuiper first described the possible existence of such small icy bodies, and we refer to the region in which they orbit as the *Kuiper Belt*. Pluto visits the inner edge of that Kuiper Belt, as do most of these other icy bodies. Pluto's existence now makes sense. It has brethren. It has a home. Pluto is a Kuiper Belt object.

Pluto is the biggest known Kuiper Belt object. That makes sense. As the first discovered object of a new species you'd expect it to be the biggest and the brightest. Ceres, which was the first asteroid discovered, is still the biggest known asteroid.

So back at the Rose Center, we simply put Pluto in an exhibit on the Kuiper Belt. Didn't even say it wasn't a planet. Physical properties mattered more to our design than labels.

And so it was—for a year, until the fateful *New York Times* article of January 21, 2001, titled "Pluto's Not a Planet? Only in New York," written by science journalist Kenneth Chang.

It hit the internet, wired news, boston.com. I spent three months of my life just fielding media inquiries.

Walt Disney's cartoon dog, Pluto, was first sketched in 1930, the same year that Clyde Tombaugh discovered the cosmic object. So the dog and the cosmic object have the same age in the American psyche. Disney is a major force in our culture. If Mercury had been demoted, no one would have cared. But it was Pluto. Who is Pluto? Pluto is Mickey Mouse's dog.

Let's complete the arguments. Reflecting on the big gap between the orbit of Mars and Jupiter in our solar system, people felt there ought to be a planet out there. That gap is too big not to have a planet. After some effort, Italian astronomer Giuseppe Piazzi, in 1801, found a planet in that gap. They named it Ceres. Everyone was excited, because a new planet had been discovered. Have you heard of that planet? No. One book from that time has the orbits of the eight known planets: Mercury, Venus, Earth, Mars, Ceres, Jupiter, Saturn, and planet Herschel (not yet renamed Uranus). Yes, Ceres (diameter 588 miles) is on the list.

Let's go to another book—30 years later: *The Elements in the Theory of Astronomy by John Hymers*—a math-intensive textbook. It lists eleven planets—Mercury, Venus, Earth, Mars, Vesta, Juno, Ceres, Pallas, Jupiter, Saturn, and Uranus—which are denoted by their symbols (♀ for Venus, ⊕ for Earth, ♂ for Mars, etc.). Neptune wasn't discovered yet. Four new planets had cropped up, all needing their own new symbols, making a total of eleven planets. Planets were things that went around the Sun. Comets also go around the Sun but were fuzzy and had tails, so we didn't call them planets.

When we found these new non-comet-like objects between Mars and Jupiter—Vesta, Juno, Ceres, and Pallas—we called them planets too. A few years later, we had found 70 more of these things. And you know what we discovered? They had more in common with one another than any one of them had with anything else in the solar system, and they were all orbiting in the same zone. We hadn't discovered new planets. We had discovered a new species of objects. Today we call them *asteroids*, a name invented by William Herschel, the man who discovered Uranus. He found that they were tiny relative to the established planets and argued that they constituted a new class of objects. Things that started out being called "planet" were later reclassified with a new name, and more importantly, we learned something new about the structure of the solar system. Our knowledge base broadened, and our understanding advanced. That all happened about 10 years after *The Elements in the Theory of Astronomy* was published.

Pluto is about 1/5 the diameter of Earth—it's smaller than the Moon, like the other Kuiper Belt objects. The four largest moons of Jupiter (discovered by Galileo when he first turned his telescope to the sky) are all larger than Pluto. Ganymede, Jupiter's largest moon, is slightly larger than the planet Mercury, but less than half as massive. Io and Europa are jostled by the other moons gravitationally and are heated by kneading due to Jupiter's tides. Io is covered with active volcanoes. Europa has an 80-kilometer-deep water ocean beneath a 10-kilometer-deep crust of ice. There is more water in the oceans of Europa than in all the oceans of Earth. Saturn's small moon Enceladus is also heated by tides and has a southern ocean below an icecap and spectacular water geysers bursting forth. Saturn's largest moon, Titan, has liquid methane lakes and a mostly nitrogen atmosphere. It rains methane on Titan, and there are frozen methane riverbeds. It shows dark features that are frozen methane-ethane regions, while its white areas are frozen water ice. Neptune's large icy moon Triton also has spectacular geysers (perhaps gushing nitrogen). Triton orbits Neptune in a backward direction and may be a captured Kuiper Belt object. As for the asteroids, Ceres, the largest asteroid, is smaller than Pluto. Vesta is next largest; rich in iron, it may have had part of its surface blasted off by an ancient collision with another asteroid. Asteroids are rocky bodies. Kuiper Belt objects are icy bodies.

In the Rose Center, we don't count planets. We say the solar system has families, and one of those families—the terrestrial planets (Mercury, Venus, Earth, Mars)—has properties in common that distinguish their members. The Asteroid Belt is another family—small rocky bodies. The gas giants (Jupiter, Saturn, Uranus, Neptune) make a family. The Kuiper Belt objects, including Pluto orbiting near their inner edge, all have similar properties. They constitute yet another family. Much further out is a large cloud of icy bodies that completely surrounds the Sun, the Oort Cloud of comets, named for Jan Oort, the astrophysicist who first hypothesized the existence of this cometary reservoir. We have divided objects orbiting the Sun into five families. That's our pedagogical paradigm. What matters is asking what properties objects have in common. A third grader can learn that the gas giants are big and low in density—it's an excuse to learn the word "density." They're big and gaseous, like beach balls. In fact Saturn's density is less than that of water. If you took a piece of Saturn and put it in your bathtub, it would float. That's cool. I knew this as a kid and always wanted a rubber Saturn to play with in the bathtub instead of a rubber duckie.

To consider Pluto a bona fide planet overlooks its fundamental properties.

In 2005, Mike Brown and his team at the California Institute of Technology discovered a Kuiper Belt object, named Eris, which was nearly identical in diameter to Pluto. Eris was originally thought to be slightly bigger than Pluto, but improved measurements in 2015 have since shown Eris (diameter: 2,326 ± 12 km) to be slightly smaller than Pluto (diameter: 2,374 ± 8 km). Eris had a small moon, Dysnomia, whose orbital parameters allowed Brown to accurately estimate Eris's mass; it is 27% more massive than Pluto. The discovery of Eris threw down the gauntlet. If Pluto was a planet, then surely Eris must be one also. You had to either demote Pluto or promote

Eris. The International Astronomical Union (IAU), the official body for deciding such definitions, held a special session in a 2006 meeting to vote on Pluto's planetary status, as well as that of Eris and the other Kuiper Belt objects. The result? Pluto was demoted to dwarf planet from its previous planetary status. The story made news around the world. Textbook writers took note. To be a planet an object had to (1) orbit the Sun, (2) be massive enough for gravity to pull it into a spherical or nearly spherical shape, and (3) have cleared the neighborhood around its orbit of debris. Pluto failed the third criterion, as did Ceres—they each share orbital zones with countless other objects, whose total mass is comparable with their own. Most astrophysicists, including Mike Brown, interpret "cleared the neighborhood" to mean that the planet must now dominate the mass in the neighborhood of its orbit. Jupiter, after all, is accompanied by more than 5,000 Trojan asteroids, clustered around gravitationally stable points either 60° ahead or 60° behind Jupiter in its orbit, but these asteroids in total are miniscule in mass in comparison with Jupiter itself. The IAU was not demoting Jupiter.

The IAU affirmed that there are eight planets in the solar system: Mercury, Venus, Earth, Mars, Jupiter, Saturn, Uranus, and Neptune. Or, My Very Excellent Mother Just Served Us Nachos. Meeting the first two criteria, Pluto, Eris, and Ceres were assigned their new classification of "dwarf planet." The Rose Center was in fact 6 years ahead of its time in demoting Pluto.

I wrote a book, *The Pluto Files: The Rise and Fall of America's Favorite Planet* (2009), chronicling my experiences. Mike Brown has written a book on his discovery of Eris, titled *How I Killed Pluto and Why It Had It Coming* (2010). We have now discovered four smaller moons orbiting Pluto, in addition to Charon. Eris has one moon, and the Kuiper Belt object Haumea (now also designated by the IAU as a dwarf planet) has two. In 2006, NASA launched the *New Horizons* spacecraft toward Pluto, its electrical power fittingly provided by radioactive plutonium it carried along. Some of Clyde Tombaugh's ashes (he died in 1997) were aboard. The spacecraft flew by Pluto and Charon in 2015, snapping beautiful pictures of them both. A heart-shaped icy region visible on Pluto has been tentatively named "Tombaugh

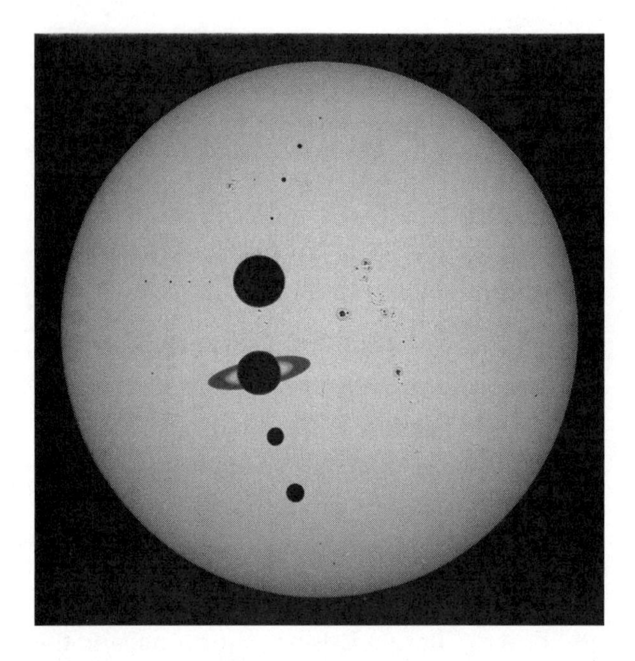

FIGURE 1. The solar system. Planets and Pluto and Eris plus some planetary moons shown as if in transit in front of the Sun. Top to bottom: Mercury, Venus, Earth, Mars, Jupiter, Saturn, Uranus, Neptune, Pluto, Eris.

Photo credit: Adapted from J. Richard Gott, Robert J. Vanderbei (*Sizing Up the Universe*, National Geographic, 2011).

Regio." So all is well with Pluto. The solar system (see figure 1) is happy: Sun, terrestrial planets, asteroids, gas giant planets, Kuiper Belt objects, and comets.

CHAPTER 3

THE LIVES AND DEATHS OF STARS

Michael A. Strauss and Neil deGrasse Tyson

Step outside on a moonless and cloudless night, in a location far from city lights, and you will see a sky filled with stars. There are roughly 6,000 stars over the entire sky bright enough to be seen with the unaided eye, half of which are above the horizon at any given time. One of the great triumphs of twentieth-century astrophysics was understanding what stars are (including our own Sun), and the full range of their properties. A star is a ball of gas, held together by its own gravity and held up by its internal pressure. We classify stars by their atomic composition (mostly hydrogen and helium, as we'll see), as well as their mass, surface temperature, size, luminosity (how much energy the star emits every second), and age. As we'll see, these various stellar properties are intimately related to one another and give us fundamental insights into the life cycles of stars, from birth to death.

Around 1910, two astronomers working independently, Ejnar Hertzsprung in Denmark and Henry Norris Russell in the United States, decided to take all the stars with well-measured properties and plot their luminosity versus their color (see figure 2). This graph is, not surprisingly, called the *Hertzsprung-Russell (HR) diagram*. The vertical axis on the Hertzsprung-Russell diagram shows luminosity, and the horizontal axis shows color or temperature, with the hottest (blue) stars on the left, and the least hot (red) stars on the right. In general, the color of a glowing ball of gas is directly related to its temperature, so the colors of stars are proxy for their surface temperature.

Henry Norris Russell, a Princeton Professor, was, by many accounts, the first American astrophysicist. A theorist, he first brought the full hammer of spectroscopic physics to the analysis

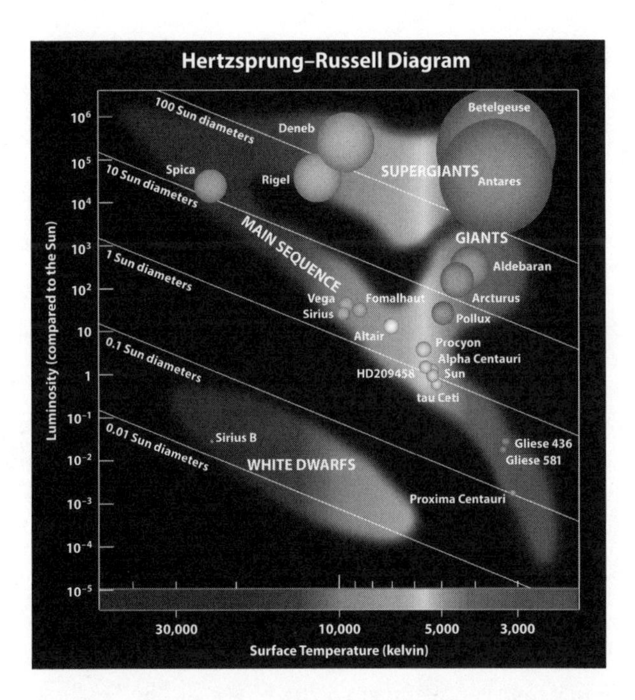

FIGURE 2. Hertzsprung-Russell diagram for stars. Luminosities of stars are plotted against their surface temperatures. Note that by convention, surface temperature decreases to the right. The shading indicates where stars are commonly found. Stars lying along a particular labeled diagonal line all have the same radii.

Credit: Adapted from J. Richard Gott, Robert J. Vanderbei (*Sizing Up the Universe*, National Geographic, 2011).

of the universe. He had data on hundreds of stars, obtained primarily by women at the Harvard College Observatory, doing what most men considered to be menial work—classifying spectra of all these stars. That was long ago, when humans who did calculations were called "computers." Yes, people were computers. The women filled one large room, where they all worked. Back then, around the turn of the twentieth century, women weren't professors and had no access to any of the jobs that men coveted. But this room of computers included smart, motivated women who, in the analyses of these spectra, deduced seminal features of the universe. Henrietta Leavitt was among them. Cecilia Payne also worked on spectra at Harvard for a decade as Harlow Shapley's assistant before eventually being appointed a professor. She discovered that the Sun is made mostly of hydrogen. Astronomy, because of that peculiar history, has a fascinating legacy of early contributions by women, detailed in Dava Sobel's 2016 book *The Glass Universe: How the Ladies of the Harvard Observatory Took the Measure of the Stars*.

From catalogs of stellar luminosities and temperatures, Hertzsprung and Russell started filling in the diagram. They discovered that stars did not occupy just any place on the chart. Some regions had no stars—the blank places in this diagram—but diagonally, right down the middle, a prominent sequence of stars emerged. They called it the *main sequence*, as is the way of our field, giving something the simplest possible name.

Understanding the HR diagram requires recognizing that the luminosity of a star depends on both its temperature and its size. You will not be surprised to learn that the hotter the star, the more energy it emits every second. Similarly, the larger the star, the higher its luminosity. This means that from measurements of a star's luminosity and temperature, you can calculate its size. In other words, a star's luminosity, temperature, and size are interdependent, forming a clean equation that allows you to calculate any one quantity from the other two. Ninety percent of the cataloged stars in the HR diagram land along the main sequence. There's a smattering in the upper right corner. These stars have relatively low temperature—which means they're mostly red. And they are highly luminous—which means they're huge. We call them *red giants*. Even higher on the upper right are the *red supergiants*. We can draw diagonal lines of constant size on the diagram: 0.01 Sun diameters, 0.1 Sun diameters, 1 Sun diameter, 10 Sun diameters, and 100 Sun diameters. Now we know how big these stars are. The Sun lies on the 1-Sun-diameter line, of course. Red supergiants are larger than 100 Sun diameters. Below the main sequence we find another group of stars. These are hot but not too hot; that makes them white. They are extremely low in luminosity, so they must be small. We call them *white dwarfs*. Some people in the UK (especially, perhaps, J.R.R. Tolkien) might prefer to write *dwarves*. But in America we form the plural of *dwarf* as *dwarfs*. Astrophysicists are not alone in their preference. Disney's 1937 film is *Snow White and the Seven Dwarfs*, not Seven Dwarves.

When the HR diagram was published, classifying stars into graphical zones, we didn't know why they were grouped that way. Maybe a star is born with high temperature and high luminosity and over time slides downward along the main sequence (simultaneously cooling and losing luminosity) as it ages—a reasonable guess, but that kind of reasoning led to an estimate for the age of the Sun of about a trillion years, much older than the age of Earth. For dozens of years, we proposed educated guesses to answer the question—until we figured out what was really going on. That insight emerged from looking at clusters of stars in the sky. The key point is that all the stars in a cluster formed at about the same time. Some clusters have a few hundred stars; others, a few thousand. Still others have hundreds of thousands of stars. If the number of stars is on the low end (like the Pleiades) we call it an *open cluster*; if the cluster is on the high end, it tends to be spherical or globe shaped, and we call it a *globular cluster*.

When you see one of these objects in the sky, it's obvious which kind of cluster you're looking at. There is no middle ground. The stars in a particular cluster have a common birthday—they formed from a gas cloud all at the same time.

The Pleiades is a young open star cluster—it's like looking at a kindergarten class. Young, bright, blue stars dominate. The Hertzsprung-Russell diagram for this cluster shows a complete main sequence and no red giants. The blue stars at the top of the main sequence are so bright that they dominate, but red stars lower down on the main sequence are also in evidence. The Pleiades shows what an ensemble of stars looks like soon after they're born, and astrophysicists estimate this cluster is only a few tens of millions of years old. From it, we can see that some stars are born having high luminosity and high temperature while other stars are born having low luminosity and low temperature—they're just born that way—along the entire main sequence.

Globular clusters show a main sequence minus an upper end, plus some red giants, which are not part of the main sequence. This is like looking at a fiftieth college reunion—all the stars are old. The red giants are the brightest and dominate the view. The cluster's main sequence still has low-luminosity, low-temperature objects, but where did the bright blue ones go? Did they exit the scene? What happened? You can probably guess where they "went": they became red giants. The upper part of the main sequence was peeling away, with luminous blue stars becoming red giants.

We also found middle-aged cases: where just part of the upper main sequence was gone and only some red giant stars had appeared.

To figure out the masses of different types of stars, we had to be clever. We measured shifts in wavelengths in the spectral lines of binary stars as they orbited each other to determine their orbital velocities—the famous Doppler shifts—and applied Newton's law of gravity. From this exercise, we discovered that the main sequence is also a mass sequence, running from massive, luminous blue stars at the top left to low-mass, low-luminosity red stars at the bottom right. Low-mass stars are born with low luminosity and low temperature, whereas high-mass stars are born with high luminosity and high temperature.

Massive blue stars on the upper main sequence live for perhaps 10 million years. That's actually not much time. Around the middle of the main sequence, a star like the Sun lives for 10 billion years, 1,000 times longer. Following the main sequence all the way down to the bottom, the low-luminosity red dwarf stars will live for trillions of years. Ninety percent of all stars are found on the main sequence simply because stars spend 90% of their lifetimes with luminosity and temperature that land them there. Think of it this way: you brush your teeth in a bathroom every day. But if we take snapshots of you during the day at random times, we're not likely to catch you in the act, because even though every day you (should) spend time brushing your teeth, you don't spend much time doing it. Some regions of the HR diagram are sparsely populated because stars are actually "passing through" those regions as their luminosity and/or temperature changes, but they do so quickly, not spending much time there. It is rare to catch stars in the act of brushing their teeth.

What's going on down in the center of stars? As you raise the temperature of a gas, the individual atoms or nuclei that make it up move faster and faster. Around 90% of the atomic nuclei in the universe are hydrogen, the same percentage found in stars. Take a blob of gas that is 90% hydrogen—it's not a star yet. Let it collapse under its own gravity and form a star. As you might suspect, the center becomes the hottest part. The centers of stars are hot enough to create a nuclear furnace that keeps the center hot. It's much less hot up on the surface. The centers of stars are so hot that all electrons are stripped entirely from their atoms, exposing their bare nuclei.

The hydrogen nucleus has one proton. When another proton approaches it, the two protons naturally repel each other. Protons are positively charged, and like-charges repel. The closer they get, the harder they repel. But now increase their temperature. Higher temperature means higher average kinetic energies, and higher velocities for the protons. Higher velocities mean that the protons can approach closer to each other before the electric forces make them turn around and back away. Turns out there is a magic temperature—about 10 million K—at which these protons are able to get so close together that a whole new short-range, *strong nuclear force* takes over, attracting them and binding them together. It's what enables *thermonuclear fusion*.

Incidentally, the strong nuclear force is also what holds more massive nuclei together. The helium nucleus has two protons and two (zero-charge) neutrons. The two protons are pushing each other apart because they are both positively charged, and it's the strong nuclear force that holds them in the nucleus. This is true for all nuclei, such as the carbon nucleus, containing six protons and six neutrons, and the oxygen nucleus, with eight protons and eight neutrons.

When two protons come together at 10 million K, just sit back and watch the fireworks. You end up with a proton and a neutron stuck together—one of the protons has spontaneously turned into a neutron and a positively charged electron, called a *positron*. And that positron is instantly ejected. That's antimatter. Exotic stuff. Positrons weigh exactly the same as electrons, but when a positron and an electron meet, they annihilate, converting all their mass into energy carried away by two packets of light called photons. This conversion follows precisely Albert Einstein's equation, $E = mc^2$, expressing the equivalence of energy and mass. In this equation c is the speed of light, 186,000 miles per second, a very high velocity that gets squared, so a small amount of mass can be converted into an enormous amount of energy—as dramatically demonstrated by atomic bombs. Also ejected in the reaction is a *neutrino*, a neutral (zero-charge) particle that interacts so weakly with other stuff in the universe that it promptly escapes from the Sun, mostly unfazed by material obstructions. Notice that charge is conserved in this reaction. We start with two positive charges (each proton has one) and end with two positive charges (on the surviving proton and on the positron). This reaction creates energy, because the total mass of the original particles exceeds total mass of the particles at the end. The lost mass was converted to energy via $E = mc^2$. What is a nucleus with a proton and a neutron? It has only one proton in it, so it is still hydrogen, but now it is a heavier version of hydrogen. We often call it "heavy hydrogen," but it also has its own name, *deuterium*.

Now I have some deuterium. Deuterium plus another proton gives me a ppn nucleus (two protons, one neutron) plus more energy. What have I just made? I now have two protons in my nucleus, and when you have two protons, it's called helium. *Helium* derives from Helios—the Greek god of the Sun. Yes, we have an element named after the Sun. That's because helium was discovered in the Sun, through spectral analysis, before we discovered it on Earth. This ppn nucleus is a lighter-than-normal version of helium, called *helium-3* because it has three nuclear particles (two protons and one neutron). Now collide two of these helium-3 nuclei: ppn + ppn and you get ppnn + p + p + more energy. This resulting ppnn is full, red-blooded *helium-4* (the stuff that fills helium balloons).

All this goes on between 10 and 15 million K in the center of the Sun, which converts 4 million tons of matter into energy every second. We would later learn that all stars in the main sequence are converting hydrogen into helium in their cores. Eventually, the hydrogen runs out, and then, all heaven breaks loose: the star's envelope expands, and it becomes a red giant. About 5 billion years from now, our Sun will become a red giant, throw off its gaseous envelope, and ultimately settle down to become a white dwarf. The galaxy offers many examples of this phenomenon. In a rare astronomical misnomer, we call these fuzzy remains of stars "planetary nebulae." Stars that are more massive will become red giants and supergiants. When the most massive among them die, they explode as supernovae, with their cores collapsing to form neutron stars or black holes.

For now, let's go back to the Hertzsprung-Russell diagram. We have the main sequence, red giants, and white dwarfs, with temperature increasing to the left and luminosity getting higher as you go upward. All stars are handed a spectral classification. A seemingly random letter that is, in fact, a relic from an early classification scheme in which they appeared in alphabetical order: O B A F G K M L T Y. Each letter designates a class of decreasing surface temperature for stars. Our Sun is spectral class G. Their approximate surface temperatures (on the Kelvin scale, which is degrees centigrade plus 273) and colors are

O (>33,000 K, blue)
B (10,000–33,000 K, blue-white)
A (7,500–10,000 K, white to blue-white)
F (6,000–7,500 K, white)
G (5,200–6,000 K, white)
K (3,700–5,200 K, orange)
M (2,000–3,700 K, red)

all of which are included in figure 2. Off to the right, beyond our chart, would be the remaining spectral classes: L (1,300–2,000 K, red), T (700–1,300 K, pale red), and Y (<700 K, deep red).

The lowest-mass stars on the main sequence (which are M stars) are less than 10% the mass of our Sun. What about L, T, and Y stars that are even lower in mass? With lower gravity, they will have lower temperature and lower density in their cores. What happens when you get a gaseous mass held together by gravity that is simply not hot enough in its center for nuclear fusion of hydrogen to take place? Such a star we call a *brown dwarf*. They're not actually brown but appear deep red and glow mainly in the infrared. Brown dwarfs in the range from about 1/80 to 1/12 of the Sun's mass feebly fuse the trace amount of deuterium that exists in their cores. Because they have fusion in their cores, we still call them stars. But objects of even lower mass—less than 13 times the mass of Jupiter (or about 1.3% that of the Sun)—will have absolutely no nuclear fusion of any sort in their cores. The planets Jupiter and Saturn are just such objects.

If you look at the temperature scale at the bottom of the figure, you can see where these spectral classes go. Spica is class B. Sirius is class A, and Procyon is class F. And Gliese 581, a star with confirmed orbiting exoplanets is class M. The Sun, class G, has exactly one solar luminosity, by definition, as can be seen by noting its luminosity on the vertical scale. This is a logarithmic scale, allowing us to plot the huge range of observed luminosities, with each ascending tick mark representing a star 10 times as luminous.

The range in luminosity among the main sequence stars in the universe is staggering. Along the top edge of figure 2 are stars with a million times the Sun's luminosity. At the bottom of the chart are stars with 1/100,000 of the Sun's luminosity. We eventually figured out that stars at the top end of the main sequence are only about 60 times the mass of the Sun, not a million times more massive. At the bottom end, they are only about a tenth the mass of the Sun but, as indicated, are much, much fainter than the Sun. So, the range of masses is large but not nearly as large as the range in luminosities.

Since luminosity is such a strong function of mass, the high-mass stars are fusing their finite store of hydrogen fuel quickly. The more massive a star is, the shorter its main sequence lifetime will be, even though it carries more fuel. Massive stars are gas guzzlers.

A 40-solar-mass star will live only 1 million years—tiny compared with a billion years. Going in the other direction, consider a star that has 1/10 the mass of the Sun. It will live about 3 trillion years, much longer than the current age of the universe—making that star very efficient in its fuel consumption. Thus, every low-mass star ever born in the universe is still fusing its hydrogen.

As noted, a star spends 90% of its life on the main sequence, fusing hydrogen into helium. But stars do other stuff in their cores during their red-giant phase. They continue fusion, beyond helium, building ever-heavier elements such as carbon and oxygen, and others down the periodic table to iron, which has 26 protons and 30 neutrons. A lot is going on during this last phase, but it happens fast, occupying a mere 10% of a star's life. Every time you bring together elements lighter than iron, to make a heavier one, the reaction loses mass, and the fusion reaction releases energy via $E = mc^2$. We say it is *exothermic*, because it gives off energy. Other nuclear processes also give off energy. Take uranium, with 92 protons. Split its nucleus into smaller ones—fission—and that's exothermic too. The Hiroshima bomb during World War II invoked uranium fission, while the Nagasaki bomb used plutonium, with 94 protons. Each of these elements has a huge nucleus. If you split them into parts, creating lighter elements, the total mass of what you get is slightly less than what you started with. That lost mass becomes the energy of these atomic bombs—dubbed "A-bombs." Most of the world's nuclear arsenal at the beginning of the Cold War consisted of A-bombs, whereas today, most of the power of our nuclear arsenal resides in bombs that fuse hydrogen into helium—"H-bombs." Just to put their relative destructive energy in perspective, H-bombs use A-bombs as their trigger, giving a sense of how devastating these fusion-based weapons really are. We know how efficiently they convert matter into energy, and that's exactly what stars do. The Sun is one big thermonuclear fusion bomb, except its awesome energy is contained by all that mass pressing down on the core. We have not yet been able to make a nuclear fusion power plant—harnessing that ten-million-degree fuel. All nuclear power plants in America, France, and other countries are contained fission power plants.

You just can't split atoms and keep getting energy forever. You can't fuse atoms and get energy forever either.

Iron, with its 26 protons and 30 neutrons, happens to be the most tightly bound nucleus on the periodic table. Fuse iron, it goes *endothermic* and absorbs energy. Fission iron, it's endothermic again. Either coming or going, there's no more nuclear energy to be released. The buck stops at iron.

Stars are in the business of making energy. If a star is cranking along, fusing its elements down the line toward heavier and heavier nuclei, and if it's getting energy for doing so, you have a happy star. That energy keeps the center of the star hot, and the thermal pressure of that hot gas keeps gravity from collapsing the star under its own weight. Let's say we have a hydrogen and helium main sequence star 10 times as massive as the Sun, converting hydrogen to helium in its core. That's scene 1. By scene 2, the core is now pure helium, but it still has a mixture of hydrogen and helium in the surrounding envelope. Fusion stops in the center, and the center can't hold the star up anymore, so what does the star do? The star's core collapses, the pressure builds, and the temperature increases, becoming even hotter—hot enough (100 million K) to fuse helium into carbon. Outside the core, hydrogen still exists and begins fusing. The region where this happens is a spherical shell. The star is now a red giant. Scene 3. We now have carbon fusing to make oxygen in the center of the carbon core in the center of the helium core in the center of the star's outer envelope, which still has hydrogen and helium. We're creating an onion of elements, layer upon layer, because it's always hottest in the middle. Each reaction releases energy. Eventually, you get iron in the middle, surrounded by successive shells of all the other lighter and lighter elements.

Therein sits the future chemical enrichment of the galaxy.

But at the moment, these elements remain locked inside a star, and they've got to get out somehow. How do we know? We're made of these elements. We also know that iron is the end of the road. If a star tries to fuse iron, doing so sucks energy out of the star, collapsing the star even faster. Stars, remember, are in the business of making energy, not absorbing it. As the core collapses faster and faster, the star implodes, leaving a tiny, superdense *neutron star* in the center, while the outer layers undergo spontaneous nuclear fusion at an explosive rate. This sudden burst of energy blows off the entire envelope and outer core of the star, causing a titanic *supernova* explosion, shining billions of times brighter than the Sun for several weeks. We'll explore the details of the supernova explosion in just a bit. But the important point for the moment is that the guts of this star are now released into the galaxy—into what we call the *interstellar medium*—chemically enriching gas clouds with heavy elements, enabling those clouds to form star systems with planets and even people.

In AD 1054, Chinese astronomers noticed a new "guest star" in the constellation we call Taurus, which was initially bright enough to be seen during the day.

Today, at that location, the expanding debris from that supernova explosion can be seen as the Crab Nebula. In the center of the Crab Nebula, a neutron star rapidly spins at 30 times a second. When a star core collapses, it retains its angular momentum and begins spinning more rapidly, like ice skaters that rotate faster by pulling in their arms. The star's magnetic field becomes compressed and intensified as well. The magnetic field at the surface of the Crab Nebula's neutron star is about a trillion times stronger than the magnetic field at Earth's surface. As the neutron star rotates, its north and south magnetic poles, which don't align with the rotation axis, swing around, forcing the neutron star to emit radio waves in two beams like a lighthouse. Every time the beam swings past Earth, we see a pulse of radio radiation. Behold, a *radio pulsar*. The first radio pulsar was discovered by University of Cambridge graduate student Jocelyn Bell in 1967.

Supernovae are rare. The last time anybody saw a supernova go off in the Milky Way was about 400 years ago, before Galileo first pointed a telescope at the heavens. That's why, in 1987, astrophysicists were particularly excited when they saw a supernova explode in the Large Magellanic Cloud, a small, nearby satellite galaxy of the Milky Way. It was bright enough to be seen with the naked eye, even though it was 150,000 light-years (1.5×10^{18} km) away. Michael was lucky enough to travel to Chile to use telescopes there for his PhD research in May 1987, just a few months after the explosion. As the Chinese did in AD 1054, he got to see this "new" star in the Large Magellanic Cloud with his very own eyes. Both Michael and Neil used telescopes in Chile for their PhD research, and they met for the first time at Cerro Tololo Observatory in the Chilean Andes.

Heavy elements up to iron are made by fusion in the cores of dying stars. While some heavier elements are made in supernova explosions, most of the rest of the naturally occurring elements, all the way to uranium, can be created in the collision of two neutron stars in a tight orbit. Such neutron-star binaries exist. In 1974, Russell Hulse and Joe Taylor discovered two neutron stars, each with a mass of 1.4 times that of the Sun, orbiting each other once every 7.75 hours. The diameter of their orbit is somewhat smaller than the diameter of the Sun. The two neutron stars are slowly inspiraling because of the energy carried away by the emission of gravitational waves, a phenomenon predicted by Einstein's theory of general relativity. For their discovery, they were awarded the 1993 Nobel Prize in physics. The two neutron stars will continue their slow death spiral toward each other until they eventually collide and merge, about 300 million years from now. On August 17, 2017, astrophysicists from the Laser Interferometer Gravitational-Wave Observatory (LIGO) observed gravitational waves from just such a neutron star/neutron star collision. Elements heavier than iron were abundant in the explosive debris. A collision like this can eject 10 Earth masses of gold. Think of it: the atoms of gold in our wedding rings were likely forged in a collision of two neutron stars billions of years ago.

Even higher-mass stars exist than the one that made the Crab Nebula. They explode too. But when one of them collapses, the increase in gravity near the center warps space so severely that it closes itself off from the rest of the universe, preventing even light from escaping. You guessed it. We call these objects black holes. A black hole is a hotel where you check in but can't check out. An *event horizon* forms to hide whatever is going on inside. A 10-solar-mass black hole has an event horizon with a radius of 30 kilometers.

On September 14, 2015, astrophysicists from LIGO witnessed the gravitational waves emitted from the collision of two black holes. The two had formed a tight binary and spiraled inward, losing energy owing to emission of gravitational waves. According to Einstein's formula $E = mc^2$ the gravitational waves carried away an energy equivalent to 3 solar masses. By studying these gravitational ripples in the geometry of spacetime, astrophysicists were able to deduce the masses of the black holes involved. The two initial black holes were 29 and 36 solar masses, respectively, while the single black hole left at the end was 62 solar masses. If you do the math, you'll notice that 3 solar masses are missing. It was converted into pure energy in a fraction of a second and emitted in the form of gravitational waves. For a tenth of a second, the luminosity associated with those gravitational waves was larger than the luminosity from all the stars in the visible universe. The discovery earned the 2017 Nobel Prize in physics.

Stephen Hawking made major discoveries about black holes. He found that owing to quantum effects that they should emit feeble amounts of thermal radiation (in the long-wavelength radio-wave region of the spectrum). This is far too feeble to be detected yet, which is why Hawking did not win a Nobel Prize, which typically demands experimental proofs. The emission of this Hawking radiation causes the black hole to slowly lose mass. A black hole of 62 solar masses will evaporate on a timescale of about 5×10^{72} years, far longer than the current age of the universe.

The Sun will also run out of hydrogen fuel in roughly 5 billion years. Its death will be less spectacular than a supernova explosion but is of particular interest to us. Analogous to the fate of higher mass stars, after the Sun has exhausted all the hydrogen in its core, consider the hydrogen shell immediately surrounding the now pure helium core. Outside the core, the shell of hydrogen has been uninvolved in nuclear fusion, because its density and temperature are simply too low. But as the core collapses, this surrounding shell of hydrogen collapses as well, becoming hotter and denser. Quickly, its density and temperature get high enough to trigger the fusion of hydrogen to helium in the shell, endowing the Sun with a renewed source of fuel to run the nuclear furnace.

Suddenly, the Sun has a new lease on life. The rate of energy production in the much larger hydrogen-fusing shell is enormous—much higher than that of the core while the star was still on the main sequence.

And so, for a brief period at least, the star produces a huge luminosity, but it takes a long time for that radiation to get out, and the increased pressure starts winning the tug-of-war with gravity. As a consequence, the outer parts of the star expand (and cool), even while the inner parts contract. The Sun becomes a *red giant*, about 200 times the current width of the Sun. About 7.4 billion years from now, we expect tidal interactions with the Sun during its red-giant phase to cause Earth to spiral into the envelope of the Sun and vaporize.

The Sun's helium core has no internal energy source, so gravity causes it to continue to contract and heat up. Only when the core temperature reaches about 100 million K will the helium nuclei start fusing to make carbon and oxygen nuclei. That helium-fusing phase will last for about 2 billion years for the Sun. That's when the supply of core helium depletes, and the core collapses once again.

For stars of the Sun's mass, we're near the end of our story. The outer parts of the star sit far away from the core, feeling only a weak gravitational pull. It takes only a bit more energy to eject these outer regions, which gently expand into space as a diffuse gaseous envelope, laying bare the hot, dense carbon-oxygen core the star left behind. It's tiny (about the size of Earth) and hot enough to appear white. So we call it a white dwarf. The white dwarf has no internal source of energy, so it slowly cools off over billions of years. By force of tradition, we still call a white dwarf a star, even though it is not undergoing nuclear fusion.

What holds up the white dwarf against further gravitational collapse? Electrons. The *Pauli exclusion principle*, named for physicist Wolfgang Pauli, states that no two electrons can occupy the same quantum state. In atoms with many electrons, the electrons stack into higher energy levels when the lower energy levels are filled. For white dwarfs, the Pauli exclusion principle means that the electrons don't like to be squeezed too close together, creating a new kind of pressure that holds up the white dwarf against gravity. Our Sun will end its life in just this way—as a white dwarf.

Let us return to stars with violent fates. Stars greater than 8 times the Sun's mass go through a much more dramatic series of reactions. There's enough mass, pressure, and temperature in their cores for carbon and oxygen to fuse into such elements as neon, silicon, and the others all the way down the periodic table to iron.

The outer layers of these more massive stars grow appreciably larger than mere red giants. They become red supergiants.

In the night sky, some bright stars are clearly red to the naked eye. Red stars that are on the main sequence have a low luminosity; none are visible to the naked eye. In contrast all the bright red stars in the sky are either red giants (like Arcturus in the constellation Boötes and Aldebaran in Taurus) or red supergiants (like Betelgeuse in Orion). Betelgeuse (pronounced "Baytlejuice") has a radius about 1,000 times as large as the Sun's and is at least 10 times the Sun's mass. In its core, helium is fusing into carbon, oxygen, and heavier elements.

Continuing our story, if the mass of the star's core is more than 1.4 solar masses, electron pressure can no longer hold it up against gravity. Compressed by gravity, the electrons merge with protons to form neutrons (releasing neutrinos in the process). This leaves us with a *neutron star*—actually a giant atomic nucleus of mostly pure neutrons. The Pauli exclusion principle holds for neutrons just as it does for electrons, and the resulting neutron pressure now supports the star against gravity. However, as neutrons are nearly 2,000 times the mass of electrons, neutron stars are tiny, only about 20 kilometers across: a neutron star can have a density of almost 10^{15} grams per cubic centimeter (or 100 million elephants per Chapstick casing, as we found back in chapter 1).

Continuing on our ascent, if a star's core is more than about twice the Sun's mass, then the neutron star that tries to form is unstable to further collapse. Gravity overcomes the neutron pressure, and a black hole is born. Whether the core collapses to form a neutron star or a black hole, the infalling material, composed of elements lighter than iron, compresses violently, triggering further nuclear reactions. The sudden release of energy ejects the entire exterior of the star, causing a supernova explosion. Stars with on the main sequence with greater than about 8 times the Sun's mass are the ones that die by exploding as supernovae, forming either neutron stars or black holes in their wake. Massive exploding stars are called *Type II* supernovae. Supernovae of *Type Ia* involve white dwarfs, although the details of how this happens are hotly debated. One model suggests that the explosion occurs when gravitational interactions in a triple-star system cause two white dwarf stars to collide.

The heating due to the collision detonates their nuclear fuel and produces a supernova. Alternatively, a red giant star in a binary system can transfer mass onto a white dwarf star, pushing it over the (Chandrasekhar) limit of 1.4 solar masses. The star then collapses, again detonating its nuclear reserves to produce a supernova.

In a supernova explosion of either type, gas ejects violently outward in all directions at up to 10% the speed of light. Most or all of the star is destroyed. Heavy elements produced in the star's core are now disgorged into the interstellar medium, ready to be included in the next generation of stars and planets.

Let us talk about those stars—the next generation. The Orion Nebula is a stellar nursery, a gas cloud already enriched with heavier elements forged in the cores of a previous generation of dying stars. In the constellation, the Orion Nebula is at the bottom of Orion's sword, hanging down from his belt. Even through binoculars, it appears markedly fuzzy, not sharp like a star. Ultraviolet light from the bright, newly born, massive O and B stars in the center of the nebula excites the surrounding gas causing it to fluoresce and show up in visible light. The Orion Nebula is actively birthing about 700 stars, many of which have disks of gas and dust surrounding them that may eventually form planets.

Other stellar nurseries are birthing thousands upon thousands of star systems. Our galaxy has about 300 billion stars, and as we'll see in the next chapter, many of them are likely surrounded by planets of their own.

How important are we in this picture? Cosmically insignificant. Our tiny planet orbits an ordinary star in a galaxy filled with hundreds of billions of stars: a depressing revelation for those who would prefer to feel large. The problem is, every time we make an argument that we're special in the cosmos—that we are in the center, that the whole universe revolves around us, that we are made of special ingredients, or that we've been around since the beginning—we learn that the opposite is true. This has been codified into the *Copernican Principle*: your location in the universe, and time you are alive, is not likely to be special. It's named for Nicholas Copernicus, who showed in 1543 that Earth was just one of several planets orbiting the Sun and therefore not at the center of the universe. In fact, we occupy a humble corner of the galaxy, which occupies its own humble corner in the universe.

Every astrophysicist lives with that reality. So should you.

CHAPTER 4

THE SEARCH FOR LIFE IN THE GALAXY

Neil deGrasse Tyson

Because we are alive, we harbor a special interest in life in the universe. In a first pass, it's not unreasonable to base questions about life in the universe on life as we know it—life on Earth. Living things all seem to have a set of properties in common. First, life as we know it requires liquid water. Second, life consumes energy—in chemical terms, we have a metabolism. And number three, life has a way to reproduce itself. I'll focus on the first, because that one yields easily to the methods and tools of astrophysics. All we need to do is to explore other worlds in search of liquid water.

Ever since the story of Goldilocks, we've known (and agreed) that things can be too hot, too cold, or just right. Take the Sun. We know it radiates energy: the closer to the Sun you are, the hotter things get, and the farther away you are, the cooler things get. If life requires liquid water, and you're too close to the Sun, the water evaporates and you get steam. Too far away? It freezes and you get ice. This leads us to conclude that an orbital swath exists—a zone—where a planet ought to sustain liquid water. Astrobiologists have named this, the *habitable zone*. All stars have habitable zones, and depending on their luminosities, the habitable zones of different stars will have different sizes, which gave us something to think about. The astrophysicist Frank Drake formalized this concept for intelligent life beyond Earth and constructed what we now call the *Drake equation*. It is not so much an equation as a way to organize our knowledge (or ignorance) about the subject. So, let us develop this equation and make a prediction, or at least understand the sources of our ignorance.

Based on everything we know about life, we think life needs a planet orbiting a star. And remembering that life on Earth evolves slowly, you need billions of years of evolution to produce intelligent life. The star has to be long lived. But not all stars live a long time. Your most massive stars are dead after 10 million years or less—not much hope for intelligent life on a planet around those stars, if what happened on Earth is any indication. We need a star that is long lived and a planet in the star's habitable zone.

If you want to converse with beings on a planet that might have life, it is not good enough for it to just have life. That life has got to be intelligent. Actually, you need more than that. Isaac Newton was intelligent, but you couldn't have a conversation with him across the galaxy. What he lacked in his day was technology enabling him to send signals across vast distances of space. The intelligent life we are looking for has to be technologically proficient at the epoch when we observe it. In other words, if it is 1,000 light-years away, it must have been technologically proficient and transmitting signals in our direction across space exactly 1,000 years ago, for that signal to reach us just now. Now imagine that technology contains the seeds for its own undoing. Suppose technology in the hands of ignorant and irresponsible people enables you to destroy yourself more efficiently than any natural catastrophe. How long before you render your own civilization extinct? That might be only a hundred years. Then, if you are looking around the galaxy, you must be lucky enough to see another planet during that hundred-year slice out of the five-billion-year history of that planet in orbit around its star, which makes the probability of finding a cosmic pen pal look really slender.

Frank Drake took all these arguments and wrote them into his equation. This formed a starting point for the search for extraterrestrial intelligence, known as SETI. He wanted to estimate the number of communicating civilizations in the galaxy we could be hearing from now: N_c. To get there, he introduced a series of fractions into one equation, with each term representing a discrete estimate based on modern astrophysics:

$$N_c = N_s \times f_{HP} \times f_L \times f_i \times f_c \times (L_c/\text{age of the galaxy}),$$

where

N_c = number of communicating civilizations we could observe in the galaxy today;
N_s = number of stars in the galaxy;
f_{HP} = fraction of stars suitable with a planet in the habitable zone;
f_L = fraction of these planets where life develops;
f_i = fraction with life that develop intelligent life;
f_c = fraction with intelligent life that develops technology to communicate over interstellar distances;
L_c = average lifetime of communicating civilizations.

Let's start with the number of stars in the Milky Way, about 300 billion. Because not every star in the galaxy would be suitable, you have to multiply by the fraction of stars that are long lived (long enough to develop intelligent life) and that also have a planet in the habitable zone (f_{HP}). That reduces the total number available to us, where we might look for intelligent life. As of the date of this publication, after heroic efforts surveying more than 150,000 stars, we have confirmed the existence of more than 4,000 exoplanets. In 1994, that number was zero.

Stars with planets turn out to be common, and many stars have multiple planets. Among stars with planets, we want to discover those with a happy planet in the habitable zone. We can find exoplanets by the gravitational tug they exert on their stars, which causes an observable jiggle in the radial velocity of the star. Planets closer in exert more of a tug, causing a larger jiggle, making it relatively easy to find planets close to their stars. But such planets will be too hot to have liquid water—not ones we want for the Drake equation. The largest survey of exoplanets to date has been conducted by NASA's Kepler satellite, named after Johannes Kepler, whose work on the orbits of planets helped inspire Newton to develop his law of gravity. Kepler (the satellite) discovered planets by measuring the slight dimming of the star that occurs when the planet passes directly between the star and your line of sight. We call this a *transit*. Jupiter's radius is 10% that of the Sun. Its cross-sectional area (πr^2) is 1% that of the Sun. When a Jupiter-sized planet transits in front of a Sun-like star, it causes a temporary 1% drop in the star's light. An Earth-sized planet, whose radius is 1% that of the Sun, will cause only a 0.01% drop in the light of a Sun-like star (see figure 1). Kepler was designed in its limit to be sensitive enough in principle to detect such a diminution in the light of a star, since its main mission was to search for Earthlike planets. Kepler indeed found many smaller planets, down to sizes comparable with Earth; most Kepler detections were Jupiter or Neptune-sized planets, which are not suitable for life as we know it.

Transits are more likely to occur when the planet orbits close to its host star, and so most of the Kepler planets discovered are too hot to support life. If the planet is far enough away to have a habitable temperature, its orbit has to line up just right for us to see it transit, and because its orbital period is longer, it makes fewer transits while Kepler is looking, lowering our chance of finding it. The Kepler satellite found only about ten confirmed exoplanets between 1 and 2 Earth diameters illuminated by an amount of radiation from its star within a factor of 4 of what Earth receives from the Sun. This number is low simply because these planets are harder to find using the transit technique, although a new NASA mission, the Transiting Exoplanet Survey Satellite (TESS) is finding more.

In spite of the challenges, one promising Kepler candidate is "Kepler 62e," one of five planets orbiting a K star located about 1,000 light-years from us. The star's surface temperature is 4,900 K. Planet Kepler 62e has a radius 1.61 times as large as Earth and receives only 20% more energy per square meter from its star than Earth does from the Sun. And it's in the habitable zone. That planet checks many boxes in our search-for-life criteria. Kepler 62e may be either a rocky planet, or an icy planet with a deep ocean covering its surface. The multiplanet system of Kepler 62 is roughly 2.5 billion years older than our solar system.

What fraction of stars (f_{HP}) are suitable with a planet in the habitable zone? G stars like the Sun make up nearly 8% of the stars in the Milky Way. We know they're okay for life, because the Sun is one of them. As already noted, stars much more luminous than the Sun exhaust their fuel too quickly to give their planets the time needed to evolve complex, intelligent life, something that required billions of years on Earth. Dimmer K stars and M stars are even longer lived than the Sun, so they fulfill that requirement nicely.

But main sequence M stars have such low luminosity that, to be in the habitable zone, the planet would have to huddle so close to the M star to keep warm that it would be *tidally locked*, with one face always pointing toward the star. Tidal forces are stronger close in. These tides force the planet into a slightly ellipsoidal shape, and its rotation is slowed till the ellipsoidal shape is locked, pointing in the direction of the parent star. (Our Moon is tidally locked, with one face always pointing toward Earth, because of just this effect.) The planet doesn't care about this, but any life on its surface just might: the side of the planet constantly facing the M star would be too hot, while the other side would be too cold. An Earthlike atmosphere would freeze out on the cold side. Atmosphere from the hot side would expand to the cold side and freeze out too, in a runaway process. Eventually all of the atmosphere would end up frozen out on the cold side, ending chances for life. The only hope for life on the planet is to have a very thick atmosphere circulating the air, reducing the extreme temperature variation from one side to the other. Such an atmosphere would have a very high pressure at its surface. Also, M stars have many more surface explosions—flares—than do stars like the Sun, which could prove fatal. These things may not make life impossible, but they do make it more difficult for life to evolve.

For these reasons, G and K stars are the best candidates, making up a respectable 20% of all the stars in the Milky Way. Given such stars, what is the chance of finding a planet in its habitable zone? After adjustment for observational selection effects, the Kepler satellite data tell us that around 10% of solar-type (G and K) stars have an Earth-sized planet with a size equal to and up to twice the diameter of Earth, receiving a stellar energy at a rate between 1/4 and 4 times what Earth receives. And if, again in accord with Kepler satellite data, about 45% of those will find themselves in the radius range they need to be habitable (having liquid water on their surface) given their reflectivity and greenhouse effect, then the fraction $f_{HP} = 20\% \times 10\% \times 45\% = 0.2 \times 0.1 \times 0.45 = 0.009$.

But a planet must satisfy other criteria as well. It must have a reasonable atmosphere. The planet's orbit can't be too elliptical—for that would move it alternately too near and too far from your star, boiling or freezing the planet's water. Fortunately, most Earth-like planets discovered by the Kepler satellite have nearly circular orbits. They often appear in multiple-planet systems, where the orbital interactions between and among the planets tend to circularize their orbits over time. The planets settle down into orbits that stay away from one another.

Researchers once presumed that binary star systems would have no planets. Since more than half the stars in the galaxy are in binaries, this would cut our fraction of candidates by a factor of 2. But the Kepler satellite has found them, orbiting far from both stars.

Three additional factors—atmosphere, noncircular orbit troubles, and binary troubles—each lower the probability of a star having a planet in a habitable zone, but combined, they probably don't lower f_{HP} by more than a factor of 2. So, let's just lower f_{HP} from 0.009 by a little to $f_{HP} \sim 0.006$.

When Drake first wrote his equation in the 1960s, we had not yet discovered any planets orbiting other stars. So f_{HP} was just anyone's guess. But now we have the data to refine our estimate, which is precisely how the equation is supposed to work.

The result $f_{HP} \sim 0.006$ is empowering. The nearest star is 4 light-years away. Go 10 times as far away, out to 40 light-years. That sphere of radius 40 light-years will have 1,000 times as much volume as a sphere of radius 4 light-years, and within that sphere you will find about 1,000 stars. With $f_{HP} \sim 0.006$ you expect, on average, to find at least six habitable planets within this radius. Yes, within our 40-light-year bubble around the

Sun, we expect to find a habitable planet orbiting another star! That means television episodes of *Star Trek* in its first season, wafting outward at the speed of light as TV signals, have probably already washed over another habitable planet with liquid water on its surface.

With the bright image of the star blocked out by a special occulting disk, designed to minimize the flooding effects of scattered starlight, a 12-meter-diameter space telescope could in principle pick out an Earthlike planet orbiting a star 40 light-years away. The James Webb Space Telescope has a segmented mirror 6.5 meters in diameter. The next generation of space telescope after that may be able to find and take pictures of an Earthlike planet in the habitable zone out to a distance of 40 light-years. The planet may be green. It may have vegetation. It may be blue. It may have oceans. We could observe the spectrum of its atmosphere and determine whether oxygen is there—a kind of biomarker, a by-product of photosynthesis and other chemical reactions that can reveal the existence of life.

If you multiply the number of stars in the galaxy (300 billion) by f_{HP} (0.006) and stop there, you have the number of planets in a habitable zone: 1.8 billion. That's huge. But not all of them matter to us. Of those in the habitable zone, we are looking specifically for the fraction f_L of those that have life. But not just any life—intelligent life. What fraction f_i of those planets with life evolve intelligent life? And what is the fraction f_c of those that develop a technology able to communicate across interstellar distances?

The last term in the Drake equation is the fraction of these civilizations that are communicating at the epoch we are observing them now. That's the fraction of the time during the age of the galaxy that they are "on." If we look randomly throughout the Milky Way, we will hit some planets that were just born, some that are middle-aged, and some that are old. The chance of catching a planet during its communicating phase at some random time during the life of the galaxy is equal to the average longevity of radio-transmitting civilizations divided by the age of the galaxy. That's a fraction too. Our last fraction.

Multiplying all these fractions together, times our original number of stars, we arrive at N_c, the number of civilizations in the galaxy from which we can receive communications—now.

And therein are the seeds of the Drake equation.

We've also discovered a loophole in the habitable-zone argument. Europa, a moon of Jupiter, has an 80-kilometer-deep ocean of liquid water covered by a 10-kilometer-thick ice sheet. Yet Europa is far outside the Sun's habitable zone. How did it get warm enough to have such an extensive ocean of liquid water? It orbits Jupiter with three other large moons that perturb its orbit gravitationally, driving it sometimes a bit closer to Jupiter and sometimes pulling it farther away. When Europa is closer to Jupiter, the tidal forces from Jupiter squeeze the poor moon into a more oblong shape. When Europa is farther away, it relaxes into a more spherical shape. This steady kneading of Europa pumps heat into it, melting the interior ice and sustaining a liquid ocean. Somebody needs to spend the money and send a probe to Europa to drill or melt its way down through the ice layer to the ocean below and try its hand at ice-fishing. See if it catches anything. If we found life forms in Europa, we might just have to call them "Europeans"!

Saturn's tidally heated moon Enceladus also harbors an ocean beneath an ice layer. So, if we had estimated the fraction f_{HP} by just counting planets directly heated by their stars, we're missing an entire category of planets with liquid water and must increase the fraction in some sensible way to account for tidally heated moons like Europa and Enceladus living far outside the habitable zone. In other words, we must broaden our concept of what a habitable zone means.

What fraction of those habitable places have life? What is f_L? Our only measure of this—our only data—comes from Earth. Biologists boast of life's diversity on Earth. But I suspect that if we find an alien, that alien will differ more from life on Earth than any two species on Earth differ from each other.

Just how diverse are we here on Earth? Line us up—it's quite a zoo. Tiny bacteria here, even tinier viruses over there, a jellyfish, a lobster, a polar bear. Suppose you've never been to Earth and someone after visiting, says frantically, "I just saw an exotic life form. It senses its prey using infrared rays. It doesn't have any arms or legs, yet it's a deadly predator that stalks its prey. You know what else? It can eat creatures 5 times bigger than its head." You promptly say, "Quit lying." But what have I just described? A snake. A snake has no arms or legs and gets along in life just fine—for a snake—stretching its jaws open, eating stuff bigger than its head.

What else? Oak trees, and people. The point is all this diversity shares one planet. And we all have common DNA, like it or not. We are all connected, chemically and biologically.

Earth is now about 4.6 billion years old. In the early solar system, debris left over from the epoch of formation wreaked havoc on planetary surfaces, because large rocks and ice balls were still raining down, depositing enormous amounts of energy. Kinetic energy was converted into heat, liquefying Earth's surface, sterilizing it. That went on for about 600 million years. If you want to see how quickly life formed, don't start 4.6

billion years ago. Instead, start about 4 billion years ago, when Earth's surface became cool enough to sustain liquid water and enable complex molecules to form. Two hundred million years later, you'll see the first evidence of life on Earth—cyanobacteria 3.8 billion years ago. The fraction of planets in the habitable zone around long-lived stars that might have life is looking pretty good, because, given the chance, our planet took only a small percentage of the total available time to make life in the first place. We still don't know exactly how this happened—it remains a biological research frontier—but top people are working on it. If making life were long and hard for nature to accomplish, maybe life would have taken a billion years, or several billion years to form on Earth. But no. It took just a couple of hundred million years, which gives us confidence that this fraction f_L in the Drake equation might be quite high, perhaps near 1.

Of course, we're limiting ourselves to life as we know it. In some circles, that reference is known by its acronym, LAWKI. We just don't know how else to think about the problem with confidence. We could write about life as we *don't* know it. But where do you start? Maybe they've got two mouths and are made of plutonium. Maybe life out there is as we don't know it, but we can't figure out how to pose the right questions. It's a practical matter, not a philosophical one. We have an example of life as we know it, that's us—it's an example of one, but it constitutes an existence proof. You're trying to prove that something exists, and you have one example of that thing staring at you on your selfie screen. The proof is already there. So, let's start with that and work our way from there. We also know that we are made out of atoms that are common in the universe.

In one episode of *Star Trek*, the original TV series, the *Enterprise* crew encounters a life form based on silicon rather than on carbon. We're carbon-based life, but silicon is also common in the universe. In the *Star Trek* episode, the silicon creature was basically a squat pile of rocks that was alive and waddled when it moved. A creative storytelling leap this was. The *Star Trek* writers were trying to broaden the paradigm of what kind of life the crew would find in the galaxy. Turns out that silicon is right below carbon on the periodic table. You might remember from chemistry class that elements in the same column all have similar outer orbital structures of their electrons. And if they have similar orbital structures, they can bond similarly with other elements. If you know that carbon-based life exists, why not imagine silicon-based life? Nothing is stopping you in principle. But in practice, carbon is about 10 times more abundant in the universe than silicon. Also, many silicon-based molecules stay tightly bound, making them unwilling players in the world of experimental chemistry that is life. At room temperature, carbon dioxide is a gas, whereas silicon dioxide is a solid (sand). We have even discovered complex, long-chain carbon-based molecules in interstellar space, such as H-C≡C-C≡C-C≡C-C≡N (with its alternating single and triple bonds). We've got acetone, $(CH_3)_2CO$; benzene, C_6H_6; acetic acid, CH_3COOH; and many other carbon-based molecules out there just floating in interstellar space. Gas clouds forged these molecules all by themselves. We see some of these molecules in comets as well; for example, Comet Lovejoy (2014) was outgassing ethyl alcohol when it passed close to the Sun. Silicon does not sustain such long-chained complex molecules and so enjoys less-interesting chemistry than does carbon. If you want to base life on a fertile kind of chemistry, carbon is your element. No doubt about it. Whatever life forms populate the galaxy, even if we don't look alike, it's a good bet our chemistry will be similar, just because of the abundance of carbon across the cosmos and its bonding properties.

Earth is our one example of life having formed in the solar system, so let's take $(f_L) \sim 0.5$. It's midway between 0 and 1, not a sure thing—a 50–50 chance. What's next? The fraction of planets in orbit around long-lived stars in the habitable zone that have life *but also have intelligence.* This doesn't look too good.

By whatever scheme you devise to measure intelligence on Earth, humans tend to sit at the top. Big brains seem to matter, and we have big brains, but elephants and whales have even bigger brains; so maybe it's not just big brains. Maybe it's a ratio. Humans have the biggest brains compared to our bodies of any animal in the animal kingdom. We're making the rules here, so we get to put ourselves at the top. But perhaps our hubris prevents us from thinking about it any other way. Let's assert that we are intelligent, and let's define intelligence as, for example, the capacity of a species to do algebra. If intelligence, which we claim we have, is defined this way, then we're the only intelligent species on Earth. Gotta love porpoises, but I'm guessing they're not doing underwater algebra, no matter how complex and thoughtful their behavior seems to be. No other species in the history of the world, besides us, has ever done algebra, so we're intelligent. For the sake of this conversation, let's define it that way. Suppose we are looking for life that we can have a conversation with. We would not use English, but some language that we presume is cosmic: that would be the language of science, the language of mathematics.

If intelligence is important for species' survival, don't you think it would have shown up more often in the fossil record? It hasn't. Just because we have it doesn't make it something really important for survival. You know, after the next global catastrophe, the roaches will likely still be here, right alongside the rats, and we will be extinct. A lot of good our brains will have done us then.

Now, maybe our intelligence gives us the chance to alter this fate. There's a Frank Cotham *New Yorker* cartoon showing two lumbering dinosaurs hanging out together, and one of them says to the other, "All I'm saying is *now* is the time to develop the technology to deflect an asteroid." Meanwhile, as we came to know, an asteroid is headed toward Earth to take them out—permanently. Perhaps we can use our intelligence to prolong the natural life expectancy of our species, by going out into space and batting asteroids out of the way before they destroy us—if we are willing to give NASA the money to do it. But that's not the only threat. There's also the threat of unforeseen emergent diseases. Look at what happened to the elm tree in America. Most of them in New England were killed off by a fungus carried by the elm bark beetle. Imagine if something like that were to attack us. A novel virulent flu virus might be all it would take to do us all in—something as contagious as COVID-19, but far more deadly.

Intelligence is no guarantee of survival. Sight, however, seems to be pretty important. Organs of sight have evolved by natural selection in many different species of animals. The human eye has nothing in common structurally with the fly's eye, which has nothing in common structurally with the eye of a sea scallop. Although there seems to be just one primordial gene for making eyes, these different kinds of eyes arose along different evolutionary paths. What about locomotion, having some way to get around? Maple trees don't have legs to run with, but they bear seeds with little wings for the wind to spread them far and wide. Locomotion seems to be important, because we see all kinds of ways of making it happen: snakes slither, lobsters walk, jellyfish use jet propulsion, bacteria use flagella. Many insects and most birds fly. People walk, run, swim, and take cars, trains, boats, airplanes, and rocket ships, so we really get around. But we are still the only ones alive on Earth doing algebra, which doesn't give me much confidence that intelligence is an inevitable consequence of the tree of life. Evolutionary biologist Stephen Jay Gould has expressed similar views. This all suggests the fraction f_i might be small. To indicate that, set $f_i < 0.1$, realizing it could be much smaller.

Once you've evolved intelligence, maybe technology becomes inevitable. One might even assume that: $f_c \sim 1$. You can do algebra, you have a curious brain, you want to make life easier, you want to have vacations, you want to watch Netflix, and so forth. With such motivations, the fraction of intelligent beings who create technology might be high. After all, the only species we know that is able to do algebra did go on to develop technology for communicating across interstellar distances. But if technology contains the seeds of its own undoing (e.g., by the invention of ever-cleverer ways of killing one another and destroying our planet), then the duration of your technological, communicating culture may be a small fraction of the age of the galaxy. As we will see in the final chapter, Rich has an argument suggesting that the average longevity of radio-transmitting civilizations is likely to be less than 12,000 years. If you divide by the age of the galaxy (~10 billion years), that is a tiny fraction.

Just for fun, and realizing how uncertain many of the terms are, let's put the numbers we have discussed into the Drake equation, and consummate the calculation:

$$N_c = N_s \times f_{HP} \times f_L \times f_i \times f_c \times (L_c/\text{Age of the Galaxy}).$$

$$N_c = 300 \text{ billion} \times (0.006) \times (0.5) \times (<0.1) \times 1 \times (<12{,}000 \text{ years}/10 \text{ billion years}).$$

$$N_c < 108.$$

Thus, we might expect to find *up to* around hundred civilizations in the galaxy communicating with radio waves now. Our biggest radio telescopes can detect versions of themselves—their extraterrestrial counterparts—all the way across the galaxy. So, we have a chance. We have only begun to search.

Besides, there are about 50 million other galaxies like ours within 2.5 billion light-years of us. Most of the galaxies in this bunch are at least

10 billion years old at the epoch we are seeing them—up to 2.5 billion years ago. If Earth is a measure of things, that's plenty of time for intelligent life to have developed within them, if it were to develop at all. That multiplies the number by 50 million, giving us possibly *up to* 5 billion extragalactic, radio-broadcasting civilizations. But these distant galaxies are far away, making signals from any of their civilizations faint compared to ones in our own galaxy. That's why people usually consider looking for extraterrestrial civilizations only in the Milky Way.

Still, searches for extragalactic civilizations should be undertaken as well.

Finally, some caveats about the Drake equation. The habitable zone may be even narrower than simple estimates indicate. If Earth were farther out than it is, it would be colder and form more polar ice; the reflectivity of Earth's surface would go up, reducing the absorption of solar energy, and Earth would get colder still. You could trigger a runaway ice age. If you were to place Earth closer to the Sun, the ice would melt, the reflectivity would go down, and Earth would get hotter still. Methane trapped in peat would be released, further adding to the greenhouse effect.

The Sun is getting hotter as it evolves over billions of years. To compensate, the greenhouse effects will have to lessen, or reflectivity will have to increase, to keep within the range of temperatures on which all of civilization is based. If a star increases in luminosity over billions of years, the habitable zone will move outward, and a planet would need to remain in the habitable zone long enough for intelligent life to develop.

Life itself can affect the balance. If the star is a main sequence M star, and evolves not much at all in 10 billion years, the planet might be habitable for simple life at the beginning, but when that life turns its carbon dioxide (CO_2) atmosphere into an oxygen-rich atmosphere, as we did, the greenhouse effect will decrease, perhaps sending it into a permanent ice age. This is another reason M stars may not be ideal for forming intelligent life.

Life can affect the habitable zone in other ways. Carbon dioxide from the atmosphere can be captured in the form of calcium carbonate in the shells of sea animals and deposited in sedimentary rock (limestone) when they die, thereby lessening the greenhouse effect. Vulcanism (volcanic activity) can pump CO_2 into the atmosphere, increasing the greenhouse effect. And, of course, life forms like humans can dig up long-buried fossil fuels like oil and coal from ancient organisms, and burn them, pumping more CO_2 into the atmosphere. Estimates of the habitable zone for a given planet therefore depend intimately on its geology, climatology, biology, and even the wisdom of its species.

CHAPTER 5

OUR MILKY WAY AND ITS SUPERMASSIVE BLACK HOLE

Michael A. Strauss

Most of the stars that you can see with your naked eye are tens, hundreds, or thousands of light-years away. Circling the sky is a faint band of light we call the Milky Way—visible if you are far from city lights. Galileo was the first to point a telescope at the Milky Way and to come up with the fundamental insight that the band of the Milky Way is composed of myriads of individual stars.

In 1785, William Herschel (who also discovered Uranus) counted the number of stars visible through his telescope in different directions to make a map of the Milky Way galaxy. He concluded that the Milky Way had a flattened lens shape, and we were located near the center.

Harvard professor Harlow Shapley took a different approach. Sprinkled around the Milky Way are about 150 globular clusters, agglomerations of up to a million stars each. In 1918, Shapley was able to estimate distances to the globular clusters and thus map their positions in three dimensions. Shapley found the center of the swarm of globular clusters was (to use the modern value) about 25,000 light-years from the Sun. The Sun was definitely off center.

The Milky Way is a flattened disk of stars about 100,000 light-years across, whose center lies 25,000 light-years from the Sun. These scales are enormous: 1 light-year is 10 trillion kilometers, so a diameter of 100,000 light-years seems incomprehensibly large. The nearest stars are about 4 light-years away, or 4×10^{13} kilometers. Thirty million Suns would have to lay side by side to reach from our Sun to the nearest star. The stars are tiny specks compared to the enormous distances between them.

Relative to its width, the Milky Way is highly flattened, only 1,000 or so light-years thick. The constellation of Sagittarius lies in the direction of the galactic center. With the dust of the interstellar medium concentrated in this disk of the Milky Way, the center of the Milky Way is heavily enshrouded by it, obscuring our view. This dust was responsible for Herschel's incorrect conclusion that we lie at the center of the Milky Way. He was unaware that left uncounted were stars that lurked behind curtains of dust in some directions. The Sun lies in the disk of the Milky Way, but if we look above and below—in directions away from the disk, like Shapley did with the globular clusters, there is little obscuration from dust, and we get a clear view, not only of those globular clusters, but of the universe beyond our galaxy.

Earth and our Sun lie close to the midplane of the Milky Way. Because the stars in the Milky Way are also largely concentrated in the flattened disk, we see the highest concentration of stars in a band that stretches in a full circle around the celestial sphere. We can see only part of that full circle above the horizon at any given time; the remainder is beneath our feet, our view of it blocked by Earth itself. In the Northern Hemisphere, we get the best view of that part of the Milky Way lying in the direction away from the center of our galaxy. Because Earth and the Sun lie far from the center, relatively few of the Milky Way's stars lie in that direction, and we get a relatively sparse view. From the Southern Hemisphere, however, one can look directly toward the heart of the Milky Way, and the view is much more dramatic, in spite of the obscuring effects of dust. On a clear moonless night in May, away from city lights, in Chile, the view is breathtaking (see figure 3). Among my fondest memories are those times I spent looking up at the sky at Cerro Tololo Inter-American Observatory in Chile next to the woman whom I would later marry, with the Milky Way dramatically splayed across the sky over our heads.

If we could somehow view the Milky Way from a vantage point a couple of hundred thousand light-years from here, where we could see it face on, it would look like the artist's conception shown in figure 4. The Sun is in a spiral arm about halfway

FIGURE 3. The Milky Way over the Cerro Tololo Inter-American Observatory in the Chilean Andes. The large dome in the center houses the 4-meter-diameter Victor Blanco Telescope. The center of the Milky Way appears near the right edge of the picture. The Large and Small Magellanic Clouds, companion galaxies to the Milky Way over 150,000 light-years away, are apparent on the left.
Photo credit: Roger Smith, AURA, NOAO, NSF.

out, directly below the center (at 6 o'clock in the diagram). Our galaxy is a *barred spiral*, because its central bulge has a bar shape. The spiral arms start from the ends of the bar.

FIGURE 4. Artist's view of the Milky Way from above.
Photo credit: NASA Chandra Satellite.

Not all stars in the Milky Way are located in the spiral arms and the bulge. Globular clusters are spread more-or-less spherically, extending above and below the disk. In addition, a sprinkling of stars, far sparser than those in the disk and also spherically distributed, extends about 50,000 light-years from the center of the Milky Way. We refer to this as the *halo* of our galaxy.

The stars in the bar, and especially in the halo, tend to be old, having formed billions of years ago. As a result, the hottest main-sequence O and B stars, with their lifetimes measured in mere millions of years, are simply not found there. Stars have not formed in the halo of our galaxy for billions of years. We find young hot stars almost exclusively in the spiral arms in the disk, where stars are forming now.

The spiral, or pinwheel, structure of the disk suggests that the whole structure is rotating. Indeed, this is exactly what is happening. The entire disk is rotating around its central axis, and the Sun in particular is moving in a roughly circular orbit with a speed of 220 kilometers per second. Just as the Sun's gravity holds Earth in its yearly orbit, so too does the collective gravity of all the stars in the Milky Way hold the Sun in a nearly circular orbit around the galactic center. The Sun goes around the Milky Way once every 250 million years. The Sun has completed approximately 18 orbits of the galaxy in the 4.6 billion Earth-years since it formed.

Knowing Newton's law of gravity, the distance to the center of the galaxy, and our speed orbiting around it, we can calculate the mass of the Milky Way within the Sun's orbital radius, and it's roughly 100 billion times the Sun's mass.

The Milky Way is made of stars, and so we can say that the Milky Way contains roughly 100 billion stars, under the crude approximation that the Sun represents an average star. But that's not quite true. The typical star in the Milky Way has a mass somewhat less than that of the Sun, and we haven't accounted for those stars farther out from the center of the Milky Way than we are, so a better estimate gives roughly 300 billion stars in the Milky Way. Carl Sagan in his classic TV series *Cosmos* often referred to "billions and billions" of stars, in his distinctive voice. Sagan wasn't exaggerating. The Milky Way indeed has billions and billions—roughly 300 billion—stars in it. We used that number in our evaluation of Drake's equation in chapter 4.

The stars in the disk are all on approximately circular orbits. Stars are like cars on a circular racetrack. The ones on the inner lanes are passing the ones on the outside lanes. The spiral pattern we see is due to traffic jams in the stars as they circle. If you are on an expressway and approach a traffic jam where the cars are going more slowly than average, you will slow down too. Eventually you pass through the traffic jam, and then you can speed up as the cars speed up around you. The traffic jam represents a *density wave* in the pattern of cars. The cars are most densely packed in the traffic jam—although individual cars are continually moving through the traffic jam and passing out of it. In the same way, a spiral density wave in the galaxy represents a gravitational traffic jam of stars, whose gravity pulls even more stars toward it. Furthermore, as the stars crowd together, the interstellar gas is pulled together by the extra gravitational force accumulated there, causing clouds of gas to collapse gravitationally and form new stars. So, the spiral arms are regions in which stars are actively forming. Among the newly formed stars are massive luminous blue stars, whose lifetimes are shorter than the time it takes for them to drift out of the traffic jam of the spiral arm. Thus, the spiral arms in galaxies are brightly lit by newly born, massive, blue stars. Stars do not travel on spiral paths—instead the spiral arms shine brightly because of star formation caused by these traffic jams of stars circling the galactic center.

The mass of 100 billion Suns that we've just estimated represents that part of the Milky Way within the Sun's orbit. The gravitational forces from different parts of the Milky Way *beyond* the Sun's orbit pull us in opposite directions: material just outside the Sun's orbit on our side of the galaxy pulls us outward, whereas stuff outside the circle of the Sun's orbit but on the other side of the galactic center pulls us inward. These opposing forces effectively cancel each other and have no net effect on the orbit of the Sun. So all that matters to the Sun are the collective sources of gravity within its galactic orbit, which act (mathematically) as if they were all located at the center.

This means we can use speeds of stars orbiting even farther out in the Milky Way to measure the mass enclosed in those larger orbits. We found that the mass interior to radius R goes up linearly with R. The farther out you go, the more mass you find. That's puzzling, because one sees fewer and fewer stars in the outer parts of the Milky

Way. That is, there is a significant component of the mass of the Milky Way outside the orbit of the Sun that is simply not visible in the form of stars or any other known objects. This matter is dark. We call it *dark matter*. We have inferred its presence solely through its gravitational effect on stellar orbits.

How much dark matter does the Milky Way contain? The vast majority of the Milky Way's mass, roughly a trillion times the mass of the Sun, is in the form of dark matter, extending perhaps 250,000 light-years from the center. We infer the same mass by calculating the mutual orbit of the Milky Way and its companion galaxy, the Andromeda galaxy, once again using Newton's law of gravity. The two were once moving apart from each other as part of the general expansion of the universe but are now falling together at a speed of about 100 kilometers per second, set to collide about 4.5 billion years from now.

Caltech astrophysicist Fritz Zwicky was the first to discover dark matter, in 1933, when he measured the total mass of the Coma Cluster of galaxies using the radius of the cluster and the velocities of the individual galaxies moving in the gravitational field of the cluster as a whole. He concluded that the cluster was significantly more massive than the total of the stars and gas making up the individual galaxies we could see. He dubbed the rest *Dunkle Materie* in his native German, that is, "dark matter." As we'll explore in the next chapter, this dark matter is almost certainly not composed of ordinary atoms but rather of elementary particles we have yet to identify.

Another interesting form of nonluminous matter in the Milky Way occurs right at its center. Infrared observations can penetrate the obscuring dust to show stars in the center of the galaxy moving on elliptical orbits. The orbits have diameters as small as 1/100 of a light-year with periods of 10 years or so. The object they are all orbiting is invisible, but again Newton's laws allow us to determine its mass: a whopping 4 million times the mass of the Sun. It is smaller than the orbits of the stars around it, and thus extraordinarily dense, and dark. It turns out to be a *supermassive black hole*, one of the universe's most fascinating class of objects. For this discovery, Reinhard Genzel and Andrea Ghez received the 2020 Nobel Prize in Physics.

Supermassive black holes power *quasars*. Gas falling straight toward a black hole will simply be swallowed and disappear without a trace, adding to its mass but otherwise having no effect. However, it is more likely that the gas has a bit of sideways motion, or angular momentum, relative to the black hole. Then, it will not fall straight in but orbit the black hole. In analogy to stars orbiting in the Milky Way, evidence shows that the gas around a black hole lies in a flattened rotating disk. The gas orbiting closest to the black hole is moving tremendously fast, at an appreciable fraction of the speed of light. It will rub against the slower moving gas orbiting a little farther out. This friction can heat the gas up tremendously, to temperatures of hundreds of millions of degrees. Hot things radiate energy. So, while the black hole itself is invisible, the gas surrounding it, before it falls all the way in, can be tremendously luminous. A quasar is a supermassive black hole, surrounded by a disk of gaseous material glowing so hot that it can outshine the entire galaxy in which it is embedded. As the gas finally enters the black hole, it adds to the black hole mass. This accretion process, operating over hundreds of millions of years, can result in black holes that reach millions, even billions, of solar masses.

The black hole in the center of the Milky Way currently has no gas falling into it now to form a disk, so it is currently quiescent and is not shining as a quasar. The black hole at the center of the Milky Way has a radius of 12 million kilometers. Venture inside that radius and you will never come back out. Once you cross that radius, it will take you only 26 seconds to hit the center, where you will likely be shredded by tidal forces. But exactly what happens inside a black hole remains a mystery, and a hotly debated topic of research. Some physicists think you may be killed just after you cross inside; others suggest that you may survive to pop out into other universes. You may even enter a time travel region trapped within the black hole, where you can meet your future self and say hello. To understand what really happens we may need to understand the laws of quantum gravity—how gravity behaves on microscopic scales. You can find out for yourself by jumping in, but you will never be able to come back out to tell your friends of your adventures.

Our study of the Milky Way has led us to the frontiers of physics, from the suggestion of new elementary particles populating the outskirts of our galaxy to a supermassive black hole lurking in its center. But the Milky Way is only one of hundreds of billions of galaxies, and we turn now to what all those galaxies can tell us about the large-scale structure of the universe.

CHAPTER 6

GALAXIES, THE EXPANDING UNIVERSE, AND THE BIG BANG

Michael A. Strauss

Stars are so far away, they appear as unresolved points of light, even through modern telescopes. But seventeenth-century astronomers noticed a number of other objects in the sky that were extended and often fuzzy looking. They gave them the generic name *nebula* (plural *nebulae*), from the Latin word for cloud. We have already encountered a variety of nebulae in this book, including planetary nebulae, which result when red giants throw off their outer layers; the Orion

Nebula, a region of intense star formation in which the surrounding gas fluoresces because of light from hot young stars; and even dark nebulae, the dust clouds that block light from background stars. There is yet another class, sensibly called *spiral nebulae*, because of their shape. The most prominent of these, the Andromeda Nebula, is bright enough to be seen with the naked eye under dark skies, far away from city lights. The astrophysicist Harlow Shapley, working at the beginning of the twentieth century, thought spiral nebulae were gas clouds within our own galaxy. The Milky Way is fuzzy looking itself. However, the spiral structure of the Milky Way disk that we discussed in the last chapter was unknown 100 years ago, because, living within the disk itself, we didn't have a good understanding of its three-dimensional structure, making it harder to detect its resemblance to that larger class of objects. Still, the German philosopher Immanuel Kant had speculated as early as 1755 that the spiral nebulae were other "island universes," that is, objects as large as the then entire known universe—the Milky Way. Given Shapley's determination of the extent of the Milky Way and the small angular size of spiral nebulae on the sky, if Kant was right, this meant that they must be astonishingly—seemingly impossibly—distant, millions or tens of millions of light-years away.

Shapley found this notion to be completely implausible. The astrophysicist who made the observations that settled the question once and for all was Edwin Hubble.

Mount Wilson Observatory, where Hubble worked, overlooking the Los Angeles basin, had the largest telescope in the world at the time—100 inches (2.5 meters) in diameter. When Hubble took pictures of the Andromeda Nebula with this telescope in the 1920s, he resolved its diffuse light into individual stars, just as Galileo had discovered when he pointed his primitive telescope at the fuzzy arc of the Milky Way, 300 years earlier. Based on repeated observations of the Andromeda Nebula, Hubble identified several stars that periodically brightened and dimmed, which he understood to be a category of stars called Cepheid variables. In 1912, Henrietta Leavitt, who worked at Harvard, had found a relation between a Cepheid's period of variability and its luminosity. Hubble was able to measure their periods, use Leavitt's relation to infer their intrinsic luminosities, and, by measuring their observed brightness, find their distances. Hubble's conclusion was stunning: the Andromeda Nebula lay at the then inconceivably large distance of almost a million light-years, putting it well beyond the known extent of the Milky Way.

Photographic images of the Andromeda Nebula, which do much better than the human eye, showed an angular diameter of 2° on the sky. Given that angular size, and Hubble's distance, its true diameter must be about 30,000 light-years. Hubble was able to infer two compelling facts: (1) the Andromeda Nebula is comparable in size to the Milky Way, and (2) Andromeda lies well beyond the boundaries of the Milky Way.

Moreover, the sky was filled with other spiral nebulae, all much smaller in angular size and fainter than Andromeda. If they were similar to the Andromeda Nebula, they must be even farther away. This was a pivotal moment in the history of our understanding of the cosmos. Kant's hypothesis that the spiral nebulae were other "island universes," as large as the Milky Way itself, was correct. The boundaries of the known universe took a dramatic leap outward.

Two decades later, astrophysicists realized there was more than one class of Cepheid variable in the sky. When everything got straightened out, it turned out that Hubble had actually *underestimated* the distance to the Andromeda Nebula. Our modern value of its distance is 2.5 million light-years. Furthermore, modern digital imaging shows Andromeda's outer fainter regions extending to a diameter of about 3° in the sky. With these larger values, we infer that the diameter of the Andromeda galaxy—yes, we now call it a galaxy—is about 130,000 light-years, somewhat larger than the Milky Way. Still, Hubble's estimate was in the right ballpark, and his conclusion that Andromeda was another galaxy like the Milky Way was correct. Even a rough estimate was good enough to answer the big question.

Not all galaxies have a flattened disk—some are rounded, elliptically shaped, and dominated by old stars, with very little gas or dust. Hubble called these *elliptical galaxies*.

Most luminous galaxies are either ellipticals or spirals, but some galaxies have irregular shapes, not fitting into either category. You guessed it. We call them *irregular galaxies*. The Large Magellanic Cloud, a small satellite galaxy (14,000 light-years across) that orbits the Milky Way at a distance of about 160,000 light-years, falls into this category. It appears at the far-left edge of figure 3, next to the observatory dome. It is so close to us that it is easily visible to the naked eye.

Currently, the Milky Way and the Andromeda galaxy are falling together under the influence of their mutual gravitational attraction. When the galaxies collide, about four and a half billion years from now, the vastness of empty space between the stars ensures that the stars will slide past one another without colliding as the galaxies pass through each other. After the galaxies settle down in a few hundred million years, we expect them to merge and become an elliptical galaxy.

The stars in elliptical galaxies tend to be older than those in spirals, suggesting that most elliptical galaxies formed earlier in the history of the universe. The bulges of spirals share many properties with elliptical galaxies, suggesting they may have formed in similar ways. Gas that falls later onto an already-formed elliptical galaxy can cool before it has had time to form stars. The cooling causes the gas to lose energy, but not angular momentum, which can make it form a thin rotating disk. This process could make a spiral galaxy containing an elliptical bulge. The details of this process are still poorly understood and hotly debated.

A very long exposure with the Hubble Space Telescope has shown 10,000 faint, distant galaxies in a tiny, dull, and boring patch of sky equivalent to 1% of the area of the full Moon, which comes to about 1/26 millionth of the whole sky. Since we have no reason to suspect that this spot is unusual, we estimate that the number of galaxies we can see in the whole sky is 26 million times as many as we can see in that patch. That means 260 billion galaxies are within the reach of the Hubble Space Telescope. Each of these barely resolved spots of light is a galaxy, as large as the

Milky Way, containing around 100 billion stars. With 10^{11} stars in each of more than 10^{11} galaxies, we infer that the observable universe contains at least 10^{22} stars. Neil boggled our minds with this number in chapter 1. Here we see where it comes from.

Astrophysicists started measuring the spectra of galaxies around 1915, a feat requiring photographic exposures many hours long. These first spectra showed features just like those seen in, G and K stars, which meant galaxies are made of stars. Edwin Hubble came to the same conclusion when he resolved individual stars in his photographic images of the Andromeda Nebula a decade later. The spectra of galaxies were refreshingly familiar to those accustomed to studying the spectra of ordinary stars. However, astrophysicists quickly noticed a significant difference. The spectral features from such elements as calcium, magnesium, and sodium were at wavelengths somewhat different from those seen in stars in the Milky Way, or in the laboratory. Typically, all the spectral lines from an individual galaxy were shifted systematically to the red. We call this phenomenon the *redshift*.

You can understand how the redshift originates by standing on a busy street corner and listening to a loud motorcycle ride by. You will hear a high-pitched whine as it comes toward you. Then as it reaches you and begins to travel away from you, the pitch of the engine's roar drops noticeably as it zips past, sounding something like "Neeeeyaooooowwww!"

Sound from the motorcycle is a pressure wave in air, which, like light, has a certain wavelength and frequency. The higher the frequency (shorter the wavelength), the higher will be the pitch your ear perceives. As the approaching motorcycle emits a succession of wave crests, they crowd together resulting in a higher pitch. Conversely, the successive wave crests that reach you as the motorcycle recedes, are stretched out by the motion and thus have a lower pitch. This effect, described by Christian Doppler in 1842, works for light waves as well as sound waves: the motion of a distant star or galaxy imprints itself as a systematic shift in the features of its spectrum.

Galaxies have specific, even unique spectral features that correspond to their precise elemental makeup. The difference between the wavelengths of these elements in a particular galaxy and those same elements seen in Earth laboratories, interpreted as a Doppler shift, tells us how fast that galaxy is moving relative to us.

By 1915, Vesto Slipher, working at the Lowell Observatory (where Pluto was later discovered) had measured the Doppler shifts of 15 galaxies. Andromeda and two other galaxies were blue-shifted, showing these galaxies were moving toward us, but all the rest—all the rest—were redshifted and thus moving away from us. The astronomical community has now measured the spectra of millions of galaxies; with only a handful of exceptions (like Andromeda), all of them show a redshift. We conclude that nearly all galaxies in the universe are moving away from the Milky Way. Are we in a special position, at the center of the motion of all the galaxies? What's really going on? Once again, Edwin Hubble made the critical measurements.

After measuring the distance to the Andromeda Nebula using Cepheid variable stars, he continued this effort with other galaxies, using a variety of ways to estimate their distances. He then made a simple plot, comparing the distances of galaxies to their speeds. What he saw was a trend: the more distant the galaxy, the higher its speed of recession. Indeed, despite the substantial measurement uncertainties, he was able to conclude that speed v and distance d were proportional to each other. The velocity v was equal to its distance d multiplied by a constant, now called the *Hubble constant* in his honor. The Hubble constant is indeed constant throughout the universe at any given point in time, but it does change with cosmic epoch. The quantity H_0 (pronounced "H-naught") represents the value of the Hubble constant at present. This proportionality between speed and distance soon became known as *Hubble's law*.

Hubble's first plot in 1929 included galaxies only out to a velocity v of about 1,000 kilometers per second, corresponding to a modern distance of about 50 million light-years. By 1931, Hubble and his colleague Milton Humason had extended the plot to include galaxies receding at 20,000 kilometers per second. That really cinched the case.

Is it really true that the Milky Way galaxy occupies a special position in the universe, a point away from which all other galaxies are moving? Such a notion would go against a recurring theme we have encountered, sometimes termed the *Copernican Principle*: that Earth is not in a special place in the universe. Ptolemy and the ancients put Earth at the center of the universe—it surely looks that way—but Copernicus demonstrated that Earth orbits the Sun. We then learned that the Sun is an ordinary main-sequence star, and although astrophysicists first thought that the Sun lay at a special place near the center of the Milky Way, Shapley's more accurate work demonstrated that the Sun lies about halfway out from the center.

Hubble's law, $v = H_0 d$, implies that the Hubble constant H_0 has units of a velocity v moving away from us (usually measured in kilometers per second) divided by a distance d, commonly measured in megaparsecs (Mpcs; i.e., millions of parsecs). A megaparsec is just 3.26 million light-years. The best estimate today is of the Hubble constant is $H_0 = 67 \pm 1$ (km/sec)/Mpc.

The Hubble expansion of the universe holds only on the scale of the distances between galaxies. Imagine pennies being taped to the surface of a balloon that is being filled with air. Galaxies, like the pennies, don't expand, rather it's the space between the pennies that's expanding. Objects held together by gravity or other forces, including individual galaxies, individual stars and planets, and even ourselves, are not expanding. In fact, even the Milky Way and the Andromeda galaxies are gravitationally bound to each other, and therefore falling together, not moving apart. But on large scales, the galaxies, like the pennies on the expanding balloon, are moving apart. The measurements of redshifts, at first glance, seem to put the Milky Way at a special place relative to the other galaxies—at the center of the expansion. But that is not the case. Indeed, from the perspective of any one of those pennies, one sees all the other pennies receding. There is no center to the expansion on the balloon's surface. There is no center to the expansion of the universe.

If the universe is expanding now, then the galaxies were closer together in the past. A galaxy at a distance *d* away from us is moving away from us with a speed of $H_0\,d$. Crudely assuming that this speed remains constant with time, we can ask, how long does it take this galaxy to travel the distance *d*? Equivalently, how long ago was that galaxy right here on top of us? The answer is its distance (*d*) divided by its speed (which is $H_0\,d$ according to Hubble's law), or at a time ($1/H_0$) ago, which works out to be 14.6 billion years ago. Turns out, the value of *d* cancels out in the equation, so no matter how far away a galaxy is away, it will hit us at the same time in the past. Extrapolating the current expansion of the universe backward into the past we find that all the galaxies are together (the pennies collide) at a single point, a beginning we call the *Big Bang*—a term coined derisively by Fred Hoyle in the late 1940s. And so Hubble's constant gives us the approximate age of the universe.

This calculation of the time since the Big Bang assumes that each galaxy moves at a constant velocity, but the modern value based on a more sophisticated calculation is close, about 13.8 billion years. This is just a bit older than the age of the oldest globular clusters, which are between 12 and 13 billion years old.

Since the universe is only 13.8 billion years old, because of the finite velocity of light, we can see out only a finite distance. Light from the most distant material we can see now has been traveling toward us at the speed of light for 13.8 billion years and has therefore traversed only 13.8 billion light-years in distance through the ever-expanding space between that material and ourselves. But we are seeing that material in the past—where it used to be. Where is it now? The expansion of the universe in the meantime has carried that material (now formed into galaxies) out to a distance of 45 billion light-years from us at the present moment. This represents the boundary of the present-day observable universe. Beyond lie other, more distant, galaxies from which we have never received photons. The space between them and us is simply expanding so fast that light from them has not had time to traverse it. So, we expect many more galaxies out there, beyond the visible edge of the universe.

In 1948 George Gamow and Ralph Alpher wondered what the universe would be like at its very earliest moments. They reasoned that the universe would be compressed near the Big Bang and would therefore be very hot and permeated with the kind of glow that hot things emit—thermal radiation. This radiation cools as the universe expands.

As it expands like a balloon, its circumference increases. Imagine a light wave circling this circumference, counterclockwise. Suppose 12 wave crests are equally spaced around the circumference of the expanding balloon like 12 race cars going around a circular race track. As the circumference of the balloon expands, the cars are all racing at the same speed, the speed of light. If they start with 1/12 of the equator separating each car from the one in front of it, they will remain equally spaced around the equator of the balloon as it expands. Each car is moving at the same speed, so a car will not be catching up with the car in front or falling behind to be rear-ended by the car behind. If the cars (imagine wave crests) remain equally spaced around the equator as the circumference of the equator of the balloon gets larger, the distance separating the cars will increase. Thus, when the circumference of the universe doubles, the distance between the wave crests of light doubles as well. This explains why light will be redshifted as the universe expands: it is a consequence of the stretching of space. This redshifting means that the hot thermal radiation in the early universe is stretched to longer wavelengths as the universe expands.

Turn the burner to high on your electric stove and it will glow red, with temperature as high as 700 K. The coils emit thermal radiation at that temperature, with peak emission at a wavelength of 4 microns (4 millionths of a meter), which sits squarely in the infrared region of the spectrum. Most of the energy emerges as heat. You can feel this infrared radiation warming your hands if you hold them above the burner. (Never touch the burner!) The spectrum of this thermal radiation was worked out by Max Planck, the father of quantum physics. Einstein, following Planck's work, showed that light comes in discrete packets of energy, proportional to the frequency of the light, called *photons*. Planck's distribution tells us that the wavelength of peak emission is inversely proportional to the temperature. As the universe expands by a factor of 10, the thermal radiation in the universe will redshift by a factor of 10 to longer wavelengths. If it starts out with radiation characteristic of a certain temperature, by the time the universe has expanded by a factor of 10 it will have cooled off to 1/10 of that temperature.

Let's start the story of the expanding universe about 1 second after the Big Bang. (Don't worry. We'll talk about what happened even earlier in the next chapter.) At 1 second, the universe was stupendously hot, about 10^{10} K (10 billion kelvins) and the density of matter was negligible compared with the energy density of all that radiation. Galaxies, stars, and planets didn't exist yet, and things were much too hot for atoms or molecules, or even atomic nuclei to form. The ordinary material of the universe at this point consisted of electrons, positrons (anti-electrons), protons, neutrons, neutrinos, and, of course, lots of thermal radiation (i.e., photons). And if, as is currently thought, dark matter consists of as-yet undiscovered elementary particles, we would expect those particles to also exist in large numbers in the universe at this time.

Two and a half minutes later, the universe has expanded and cooled to a temperature of "only" a billion kelvins. At this time, the photons' thermal spectrum peaks in gamma rays. A billion kelvins is cool enough to permit nuclear fusion reactions in which neutrons and protons manage to stick together. In the Sun, we found that under high temperatures and densities, protons fuse to make helium nuclei. In the center of stars like the Sun, it takes billions of years to turn 10% of the hydrogen into helium. The nuclear reactions taking place in the early universe go much faster, because free neutrons as well as protons are present. As we learned previously, proton-proton collisions require high energy, because both protons are positively charged and they repel each other, making actual collisions infrequent. Neutrons are electrically neutral and are not repelled by protons, so neutron-proton collisions occur more often. Fusion can occur by adding neutrons to protons, on the way to producing helium. This allows the slow first steps of the solar fusion process (in which protons collide to form the nuclei of deuterium atoms) to be skipped.

Protons and neutrons can transmute into each other. A neutron plus a positron can combine to give a proton plus an anti-neutrino, and vice versa. A neutron plus a neutrino can combine to give a proton plus an electron, and vice versa. And a neutron can decay into a proton by emitting an electron and an anti-neutrino. When the universe is about 1 second old, we've seen that the temperature was about 10 billion kelvins, and these processes are roughly in balance. But by the time the universe cools to a billion kelvins as it continues to expand, this balance changes so that more neutrons are converted into lighter protons, yielding seven protons for every neutron. At this point, the universe has cooled enough for a neutron and a proton to collide and stick together to form a *deuteron*, the nucleus of heavy hydrogen—deuterium. A deuteron can then participate in additional nuclear reactions to form a helium nucleus (two neutrons and two protons). After just a few minutes of nuclear fusion, nearly every neutron is incorporated into a helium nucleus, and by that time, the universe has cooled and thinned out enough that these nuclear reactions stop.

How many helium nuclei result? There are two neutrons in each helium nucleus. With a ratio of one neutron for every seven protons, those two neutrons are paired with 14 protons. Two of those protons are also included in the helium nucleus, with 12 protons left over. This predicts that one helium nucleus forms for every 12 protons (these, of course, are hydrogen nuclei). After these first few minutes, the universe's density and temperature have dropped to the point that no further nuclear reactions can take place. Thus, a significant number of helium nuclei are made in the Big Bang, along with trace amounts of leftover deuterons, lithium, and beryllium nuclei (which decay into lithium), and no heavier elements.

This basic calculation, complicated yet straightforward, was first done by George Gamow and Ralph Alpher in 1948. One helium nucleus for every 12 hydrogen nuclei is excellent agreement with the results, dating back to the work of Cecilia Payne-Gaposchkin, showing that stars are composed of about 90% hydrogen and 8% helium. Thus, our predictions for the conditions in the universe just a few minutes after the Big Bang have given us a basic explanation of why hydrogen and helium are the two most abundant elements in the universe, and why they exist in the proportions we see. This astonishing success of the Big Bang model gives us strong justification for extrapolating the expansion of the universe backward in time to just a few minutes after the Big Bang.

Gamow and Alpher hoped originally to explain the origin of all the elements from the Big Bang, but their calculations showed that the nuclear reactions proceeded only through the lightest elements. All the heavy elements (including the carbon, nitrogen, and oxygen in our bodies, and nickel, iron, and silicon, which contribute to the makeup of Earth) were created later by nuclear processes taking place in the cores of stars, as described in chapter 4. Fred Hoyle, a rival of Gamow's, hoped to demonstrate just the opposite: that both the heavy and the light elements could be created from hydrogen by nuclear cooking in the cores of stars without invoking an early hot dense phase in the universe's history, and he spent much of his career trying to do so. He developed much of our modern understanding of the formation of the heavy elements in stars. But the quantity of helium that stars make is not nearly enough to explain the amount that we observe.

The fact that we see some deuterium in the universe today also points to a Big Bang origin. Deuterium is fragile and is destroyed by being fused into helium in the cores of stars, rather than being manufactured there. Stars can't make it. The only way we know it can be made is in the Big Bang, and the amount of deuterium created in those first few minutes—one deuterium for every 40,000 ordinary hydrogen nuclei—is in excellent accord with the observed value. To Gamow, the observed cosmic abundance of deuterium was a smoking gun pointing to the Big Bang.

As the universe continues to cool, its makeup does not change for about 380,000 years.

Up to that point, the material of the universe is a *plasma*, just like the glowing interiors of stars. In a plasma, the atomic nuclei and the electrons are not bound together but move independently of each other. If an electron is briefly captured by a proton, forming an atom of neutral hydrogen, it will quickly get hit by one of the many fast-moving, high-energy photons present, kicking the electron free of the proton. Moreover, because photons interact so strongly with free-flying electrons, a photon can't travel very far before it collides with another electron and *scatters* off in a different direction. That is to say, the universe at that time was opaque: like a thick fog in which you can't see very far in front of you.

The story changes drastically when the temperature drops below 3,000 K, at a time we call *recombination*, about 380,000 years after the Big Bang. At this point, the photons no longer have enough energy to keep protons and electrons from combining, at which point they pair up to make neutral atoms. Neutral hydrogen does not scatter photons nearly as much as individual free electrons do, and the universe suddenly becomes transparent: the fog has lifted. The photons can now travel on straight paths across most of the universe.

This suggests that we, in the present-day, should be able to see those photons, which have been streaming freely toward us ever since the universe became transparent, 380,000 years after the Big Bang. These photons should come from every direction in the sky. That is, in every direction we look, there is material at the appropriate distance such that the photons it emitted 380,000 years after the Big Bang are just reaching us today.

These photons are emitted by gas at a temperature of 3,000 K and so should have a thermal spectrum appropriate to that temperature. The intensity of this spectrum peaks at a wavelength of about 1 micron (10^{-6} meter), which corresponds to the infrared. But we must remember that the universe is expanding. So this 3,000 K thermal spectrum is redshifted. The universe has expanded by a factor of about 1,000 from when it was 380,000 years old until today, 13.8 billion years later. The wavelength of the radiation is stretched by the same factor that space expands. As a result, the peak wavelength of the thermal radiation now is 1 millimeter—microwaves. If the peak wavelength has increased by a factor of 1,000, then the temperature has decreased by that same factor, which means today we should see this thermal radiation with a temperature of about 3 K coming to us from all directions in the sky.

In 1948, Alpher and Robert Herman, both Gamow's students, predicted that the universe today should still be filled with this thermal radiation left over from the Big Bang, and they calculated that by now its temperature should have dropped to about 5 K.

By the 1960s, the Herman and Alpher prediction was largely forgotten, until 1965, when two Bell Labs scientists, Arno Penzias and Robert Wilson, discovered the cosmic microwave background (CMB) radiation that Herman and Alpher had predicted. Penzias and Wilson reported a temperature of 3.5 K in their original paper (later refined with more accurate measurements to be 2.725 K). The discovery of the CMB convinced the astronomical community that the Big Bang model was correct. Gamow and Alpher's prediction of the existence of the CMB and Alpher and Herman's estimate of its current temperature as approximately 5 K together constitute one of the most remarkable predictions in the history of science to be subsequently verified. Penzias and Wilson were awarded the 1978 Nobel Prize in Physics for their discovery.

David Wilkinson (my scientific grandfather—the PhD thesis advisor of my PhD advisor) was one of the scientific leaders of the NASA Cosmic Background Explorer (COBE) satellite, designed to measure the CMB spectrum to high accuracy. It succeeded spectacularly. The CMB spectrum that COBE measured follows Planck's formula for thermal radiation precisely.

Another big question that Wilkinson tackled: how uniform is the CMB—that is, does it have the same intensity (or equivalently, the same temperature) in all directions?

A perfectly smooth universe will expand uniformly, and no structure will ever form: no galaxies, no stars, no planets, no humans to look up at the sky and wonder what it all means. The fact that we live in a universe with structure, with real deviations from uniformity, tells us that the early universe, and thus the CMB, could not have been perfectly smooth.

How did structure form in the universe? Consider a region in the early universe in which the density of matter is slightly higher than in the neighboring regions. The mass associated with that region is also slightly higher, and thus it has a slightly higher gravitational pull than the material around it. A random hydrogen atom or particle of dark matter will be attracted toward that region, thereby increasing its density at the expense of the regions around it. Material falls toward this region, increasing its mass, and in the gravitational tug-of-war, it will be even more effective in pulling extra matter toward it. As time goes on, this process will cause small fluctuations in the density of matter to grow with time—enough to form the structures we find around us today. Our colleague in Princeton's Physics Department Jim Peebles has a wonderfully succinct way to describe this process of *gravitational instability*: "Gravity sucks!" he likes to say.

Given the amount of structure we observe in the universe today, and the physics of gravity, how strong should the fluctuations in the early universe be? And consequently, how large are the expected undulations in the CMB? It's a tricky calculation. You have to understand all the components of matter, both dark matter and the ordinary stuff made of atoms. We mentioned that when the universe was completely ionized (before 380,000 years after the Big Bang), photons continually scattered off the free electrons in the universe. The pressure from those photons kept fluctuations in the distribution of ordinary matter (electrons and protons) from growing under gravity. If this were the full story, the fluctuations could have been growing via gravity only since the time the universe became neutral, and the nonuniformities in the CMB would have had to have been one part in 10,000, which was not observed.

However, as Jim Peebles realized in the 1980s, dark matter can explain the discrepancy. Dark matter is *dark*; meaning it does not interact with photons, and therefore fluctuations in the dark matter can grow under gravity impervious to the pressure of photons. After the universe becomes neutral, ordinary matter can fall into lumps of dark matter that had already been growing for some time. So, if there is dark matter, we can start off with fluctuations in the CMB that are smaller than if there were only ordinary matter present: around one part in 100,000. The fluctuations in the CMB that had to be there (according to our understanding of growth of structure in a hot Big Bang universe) were finally detected by the COBE satellite, at a level of one part in 100,000, just the level Peebles and others invoking dark matter had predicted. These discoveries solidified our understanding of the expanding universe. In recognition of their fundamental contributions to this work, George Smoot and John Mather from the COBE team won the 2006 Nobel Prize in Physics, and Jim Peebles won the 2019 Nobel Prize in Physics.

So, the dark matter inferred from the rotation of galaxies is also needed to understand the CMB.

What is dark matter made of? Detailed comparison of the abundance of helium and especially deuterium with the predictions that come from the early universe tell us that the average density of ordinary, familiar matter—protons, neutrons, and electrons—is a mere 4×10^{-31} grams in each cubic centimeter of the universe. But the motions of galaxies tell us that the total density of matter in the universe is roughly 6 times larger. The difference is due to the presence of dark matter, which therefore cannot be made of ordinary protons, neutrons, and electrons.

We suspect that dark matter is composed of unseen elementary particles of a yet-to-be-discovered type, presumably forged in the extreme heat and pressure of the early universe, just as protons, neutrons, and electrons were. There are a number of speculations as to what these elementary particles might be. The theory of *supersymmetry* predicts that each particle we observe should have a massive supersymmetric partner: the *photino* for the photon, the *selectron* for the electron, the *gravitino* for the *graviton,* and so forth. The search is on at the Large

Hadron Collider for such particles. In 1982, Jim Peebles proposed that dark matter is composed of weakly interacting massive particles (and yes, astrophysicists actually acronymed that to "WIMPS") considerably more massive than the proton. Indeed, the lightest supersymmetric partner of a known particle might just fill the bill.

A detailed comparison between the statistics of the pattern of fluctuations seen in the CMB shows astonishing agreement, not only among three dedicated missions—COBE, WMAP (Wilkinson Microwave Anisotropy Probe), and the European "Planck" satellites—but also with theoretical calculation based on Big Bang physics, that includes the effects of dark matter, plus inflation and dark energy, which we will learn about in chapter 7.

After recombination, material starts gathering into ever-denser lumps to make the first stars and galaxies. But given the angular size of structures we see in the CMB, we predict that there should be substantial structure in the universe on larger scales than just galaxies, themselves a mere

100,000 light-years across. That is, the galaxies should not be randomly distributed in space but should be organized into even larger structures. To map these structures, we return to Hubble's law. When we look at an astronomical image, we see objects as if painted on the two-dimensional dome of the sky. We have no depth perception at all. But Hubble's law gives us a method to explore the third dimension: by measuring the redshift of each galaxy, we can determine its distance and see where the galaxies are located in space.

Starting in earnest in the late 1970s, astrophysicists began measuring the redshifts of thousands of galaxies to make three-dimensional maps of their distribution. They immediately noticed that the locations of galaxies are not at all random. They found clusters containing thousands of galaxies, up to 3 million light-years across, as well as voids—300 million light-years across—almost completely devoid of galaxies. The Sloan Digital Sky Survey has measured redshifts for more than 2 million galaxies. Figure 5 is a map of a small fraction of these galaxies, those in a 4° slice in Earth's equatorial plane.

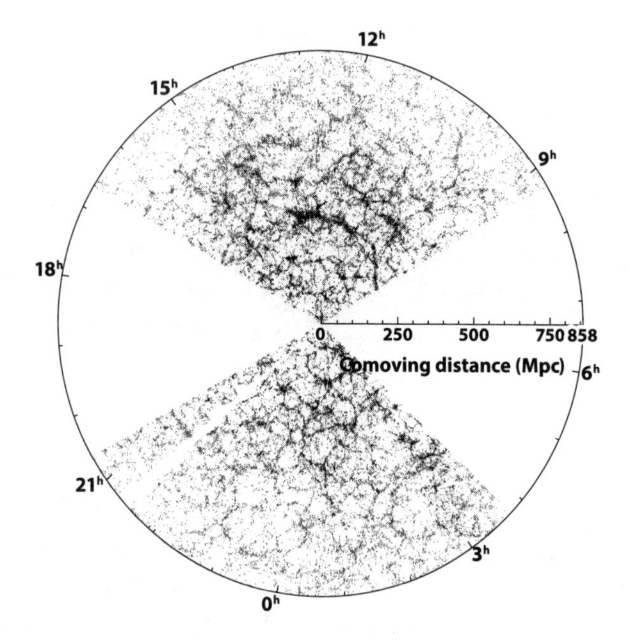

FIGURE 5. Distribution of galaxies in an equatorial slice from the Sloan Digital Sky Survey. The Milky Way is at the center. Each dot represents a galaxy. The two fans show galaxies in the survey region; the two blank regions are regions the survey did not cover. The radius of this diagram is about 2.8 billion light-years.

Credit: J. Richard Gott, M. Jurić, et al., 2005, *The Astrophysical Journal*, 624, 463–84.

Each of the more than 50,000 dots in this figure represents a galaxy of 100 billion stars.

We can see two big slices of the pie; the Milky Way galaxy sits at the center of the image. The empty regions on the left and the right fall outside the survey region, where dust from the Milky Way makes it difficult to pick out distant galaxies.

The radius of this figure is 860 megaparsecs, almost 3 billion light-years. Even a cluster of galaxies appears small in this picture, but the majority them lie along filaments, hundreds of millions of light-years long. A particularly prominent filament with a length of 1.37 billion light-years, dubbed the Sloan Great Wall, appears somewhat above the center of the image. Measuring this structure got Rich and his colleague Mario Jurić into the 2006 *Guinness Book of World Records* for "largest structure in the universe."

Notice in the figure that the density of galaxies drops off near the outer edges of the map. That's because galaxies in these regions are the most distant from us and are therefore the faintest. Only a small fraction of the most distant galaxies are bright enough for the Sloan Digital Sky Survey to measure their redshifts so that they can be included in this map.

Just as with the CMB, we can compare the patterns of structures traced by the galaxies with what we'd expect in an expanding universe subject to gravity. The result? Detailed computer simulations, including the effects of dark matter and dark energy, give rise to structures whose statistical properties beautifully match those seen in the Sloan image.

This, then, is the final triumph of the Big Bang model. We have explored the predictions of the model and compared them with observations in every way that we could. We inferred that the universe was born 13.8 billion years ago, slightly older than the ages of the oldest stars. We concluded that hydrogen and helium nuclei were formed in the first few minutes after the Big Bang, in a 12:1 ratio, which is exactly what we observe. We predicted the existence and the temperature of the CMB. We predicted that the CMB should not be perfectly smooth but should show fluctuations at one part in 100,000, with statistical properties in beautiful accord with observations.

Finally, computer models of how these fluctuations should grow under gravitational instability predict a highly structured universe today, with galaxies arrayed along filaments hundreds of millions of light-years long, just as the maps from the Sloan Digital Sky Survey reveal. The Big Bang model is far more than "just a theory." It's supported by a vast array of empirical, quantitative evidence and has passed every test we have given it with flying colors.

CHAPTER 7

INFLATION AND THE MULTIVERSE

J. Richard Gott

We interpret Hubble's results on the expansion of the universe in terms of Einstein's theory of general relativity, his theory of curved spacetime to explain gravity. In 1915, Einstein published the field equations of his theory. Newton's theory of gravity described masses creating forces that acted at a distance. Newton's law of gravity said two bodies attracted each other with a force that is proportional to the product of their masses divided by the square of the distance between their centers. But Einstein explained gravity in a completely different way. He had already shown in his theory of special relativity that we lived in a 4-dimensional universe with one dimension of time and three dimensions of space (height, width, and depth). This theory led to his famous equation $E = mc^2$. Einstein's theory of general relativity pointed out that this 4-dimensional *spacetime*, as he called it, could be curved, and his equations showed how. Masses then simply move along the straightest possible trajectories in this curved spacetime. On Earth's curved surface an airplane flying straight ahead will fly on a great circle route, called a *geodesic*. A plane flying the shortest distance from New York to Tokyo—along a great circle route—will pass over northern Alaska. This will look curved on a traditional and familiar Mercator map of the Earth, but actually it is the straightest path on the globe. Very simple in concept, and moreover, there is no action at a distance. The stuff of the universe (matter, radiation) at a location causes spacetime to curve in a specific way *at that location*. Particles, and planets, also get their marching orders locally: they just go straight ahead in the curved spacetime. The math required to derive the equations for curved spacetime is difficult, however, and it took Einstein 8 years of hard work, with help from others, to figure out how to do it. Einstein's equations were far-reaching. They even explained a gradual rotation, or precession, in the orbit of Mercury, which astrophysicists had observed but Newton's law of gravity could not explain.

Einstein calculated the geodesic path that light would take in the curved empty spacetime around the Sun. He found that a light beam from a distant star passing near the edge of the Sun on its way to Earth would be deflected by 1.75 seconds of arc. This was twice the amount Newton would have calculated, 0.875 seconds of arc. How do you observe stars near the Sun? Just wait for a total solar eclipse, when the Moon passes directly between Earth and the Sun, blocking out entirely the bright light from the solar surface. To see these deflections, you could measure the exact locations of stars on a photograph taken during the eclipse, and compare with positions you had measured many months earlier when the Sun was far away from those stars on the sky. Einstein himself proposed this as a dramatic test of his theory.

Two British expeditions were mounted to observe the solar eclipse occurring on May 29, 1919. One to Sobral, Brazil, and the other to Príncipe Island off the west coast of Africa. Sir Arthur Eddington reported the results at the combined Royal Society and Royal Astronomical Society meeting in London on November 6, 1919. From Sobral a deflection of 1.98 ± 0.30 seconds of arc was observed, while from Príncipe a deflection of 1.61 ± 0.30 seconds of arc was observed. Both results agreed with Einstein's value of 1.75 seconds of arc to within the observational uncertainties of ±0.30 seconds of arc, and both disagreed with Newton. The physicist J. J. Thompson, discoverer of the electron, chaired the meeting and pronounced general relativity to be "one of the highest achievements of human thought."

At the close of the twentieth century, there was a television program on that century's greatest moments in sports. It featured, as you might expect, Jesse Owens winning the 100-meter dash at the 1936 Berlin Olympics; Secretariat winning the Belmont Stakes by 31 lengths, to complete horse racing's Triple Crown; and Muhammad Ali knocking out George Foreman in Zaire, to regain the world heavyweight boxing championship. If that program had included the greatest plays in twentieth century science, it surely would have imagined Newton and Einstein on a basketball court, with Newton dribbling the ball downcourt. And it's not just any ball, it's his theory of gravity—the proudest thing he ever did. Einstein comes along, steals the ball, shoots it up, and, *swish*. It's the greatest play in science in the twentieth century.

In 1922, Russian physicist Alexander Friedmann found a dynamical solution to Einstein's field equations of 1915 that impacted all of cosmology. Recall our image of a balloon with pennies representing galaxies pasted on its surface. Friedmann's hypothetical balloon starts off with zero size. That is the moment of the *Big Bang*. Inflate the balloon. As the balloon expands the pennies move apart. Sit on one of the pennies. A penny close to you will be moving slowly away as the balloon expands. A penny farther away will be moving away from you faster. In fact, a penny twice as far away will be moving away from you twice as fast. Three times as far away, 3 times as fast. Well, that's precisely Hubble's law. The farther away from you another penny is, the more expanding rubber is between you and it, and the faster away from you it moves. The other galaxies are moving away from us because the curved space we live in is expanding. No penny is special—the universe looks the same as seen from any penny. The Copernican Principle is obeyed. In this metaphor we are showing only two dimensions of space, the two dimensions of the balloon's surface. That's the entire representation of our universe. Forget the inside and outside of the balloon. They don't matter. Friedmann's model actually possesses all three dimensions of space, expanding as the surface of a higher dimensional volume called a 3-sphere, but any two-dimensional slice through Friedmann's universe looks exactly like the expanding surface of a balloon. A balloon illustration is accurate, but leaves out one dimension of space so that our brains can visualize what's going on. In Friedmann's model, the balloon continues to expand but slows down and eventually stops as it reaches its maximum size, and then begins to contract, finally becoming a point at the end, in what is called the *Big Crunch*.

A *spacetime diagram* of this (see figure 6) looks like a vertical football, positioned as it would appear when ready for kickoff.

Time goes vertically in this diagram, with the future toward the top. Only one dimension of space is shown as a circular cross-section (of the balloon) whose radius changes with time. The only thing that is real here is the “pigskin” surface itself. The inside and outside of the football are not real. We are embedding the football in a higher dimensional space only so that we can visualize it. The balloon universe starts with zero radius at the Big Bang (at the bottom). Then it expands to larger circumference with time until it reaches a maximum circumference in the middle of the football, and then begins to shrink, finally collapsing to zero radius at a Big Crunch at the end. Galaxy paths through spacetime, called their *worldlines*, are geodesics along the seams in the football starting at the Big Bang and ending at

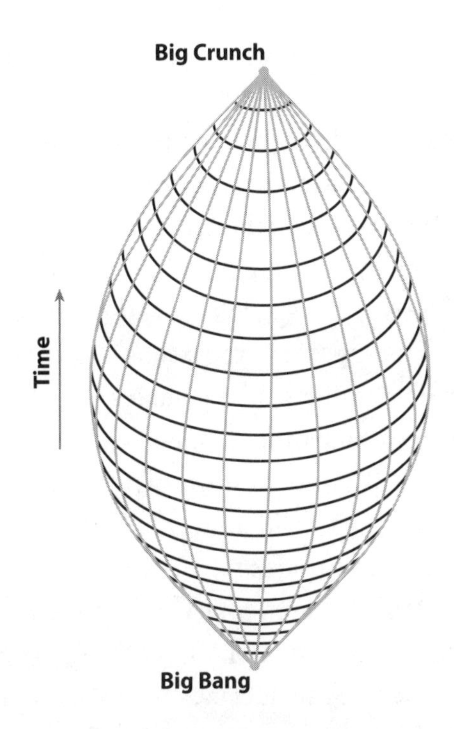

FIGURE 6. Friedmann Big Bang universe. This spacetime diagram shows only one dimension of space (the circumference of the football shape) and one dimension of time (vertical). The Big Bang is at the beginning; the universe expands at first but ultimately contracts, ending in a Big Crunch.

Credit: Adapted from J. Richard Gott (*Time Travel in Einstein's Universe*, Houghton Mifflin, 2001).

the Big Crunch. These worldlines are as straight as possible. You could drive a little truck along them and not have to turn your steering wheel. This shows Einstein's equations working at their best. The mass of all the galaxies in the universe is causing the spacetime to be curved, and the curvature of the spacetime causes the worldlines of the galaxies—the seams—to bend. Just after the Big Bang, the galaxies are all flying apart from each other. We currently live in the bottom half of the football, when the galaxies are moving apart with time. But in this example, gravitational attraction (curvature) slows their expansion to a momentary halt in the middle, at the football's equator, and then, in the upper half of the football, it causes the galaxies to start moving toward one another. The distances between galaxies, as they follow the seams, start decreasing as the circumference of the universe begins to shrink. They all crash together at the Big Crunch. You don't want to be around then! As the volume of the universe shrinks to zero, you will be crushed as you hit a Big Crunch *singularity*, a place where the curvature becomes infinite.

Time begins at the Big Bang—a singularity of infinite curvature is there. What happened before the Big Bang? This question makes no sense in the context of general relativity, because time and space were created at the Big Bang. It's like asking what is south of the South Pole. If you go farther and farther south, you will eventually get to the South Pole. But you can't venture any farther south. Likewise, if you go farther and farther back in time, you will eventually get to the Big Bang. That's when time and space were created, so that's the earliest you can go.

Friedmann's 1922 model represented a high-density universe, where the gravitational attraction of the galaxies for one another was strong enough to eventually halt the expansion and start a contraction ending with a Big Crunch. This universe contains a finite number of galaxies. In 1924, Friedmann produced a model for a low-density universe. Here, the gravitational attraction of the galaxies for each other is not large enough ever to overcome the expansion and the universe expands forever. We will return to consider this low-density model later.

Although Friedmann published his high-density universe solution in 1922, almost nobody paid any attention to it. Not until Hubble discovered the expanding universe in 1929. The galaxy worldlines were moving apart. That would put us in the lower half of Friedmann's 1922 vertical football, during the expansion phase.

Remember, no one had talked about anything like an expanding universe before. People would have asked: Expanding into *what*? But in Einstein's theory, it's curved space itself that would be expanding. It's not expanding into anything; it's just stretching. It's the space connecting all the galaxies that is just getting bigger. Amazing. It was outside-the-box thinking and a theoretical leap. *Time* magazine would later name Einstein the most influential person of the twentieth century.

The Friedmann Big Bang model has been remarkably successful (see chapter 6), but some questions remained. Why was the Big Bang so uniform? When we look out in different directions, the temperature of the cosmic microwave background (CMB) radiation is uniform to one part in 100,000. How do these different regions "know" to be at the same temperature? When we look in one direction, we see out 13.8 billion light-years. We are looking back to a time when the universe was only 380,000 years old. If we look out 13.8 billion light-years in the opposite direction, 180° across the sky, we see another region that is at essentially the same temperature. In the standard Big Bang model, these two regions on opposite sides of the sky, 380,000 years after the Big Bang were separated by a distance of 86 million light-years. That's too far away for any means of influence (even via the speed of light) to have propagated during the brief time the universe had been around. Usually, if we see two regions at the same temperature, it is because they have had time to interact, share energy, and reach thermal equilibrium. But in the standard Big Bang model, widely separated parts of the CMB have not had time to be in causal contact with each other. If so, how could it be so uniform?

And where did the fluctuations of one part in 100,000 come from? The universe was almost perfectly uniform, but not quite. It was a mystery. We needed to explain the overall uniformity and then the minute fluctuations as well.

In 1981, Alan Guth proposed a solution: the universe started with a short period of accelerated expansion he called *inflation.* In a spacetime diagram, this looks like a small trumpet bell pointing upward to hold up the Friedmann football spacetime (see figure 7).

This model starts with a finite circumference near the mouthpiece of the trumpet that then becomes dramatically larger as we move upward in time to the bell-like opening of the horn. The bottom tip of the Friedmann football is replaced by a little trumpet mouth with finite circumference at the bottom—perhaps as small as 10^{-26} centimeters. The trumpet epoch lasts a little longer than the Big Bang tip of the football would alone, and this extra time allows the different regions we see in the CMB today enough time to come into causal contact. In the beginning, the circumference is so small that the different regions, benefiting from that little bit of extra time, can come into causal contact, and then the

FIGURE 7. Inflationary beginning (trumpet) to start a Friedmann Big Bang universe (football).

Photo credit: J. Richard Gott.

accelerated expansion during the trumpet epoch pulls them far apart.

What caused the inflation if that's what really went on in the early universe?

In 1917, in a failed attempt to produce a cosmological model in which the universe is static—neither expanding nor contracting—Einstein introduced a new term into his field equations called the *cosmological constant*. Friedmann dropped the cosmological constant and found the dynamical Big Bang cosmology we described above. But in 1934 Georges Lemaître showed that incorporating the cosmological constant was equivalent to assuming that the universe had a small vacuum energy density accompanied by a negative pressure (a sort of universal suction) of equal magnitude. We are used to thinking that empty space has a zero density. It has, after all, been cleared of all particles and radiation. But the vacuum of empty space may have an energy density due to fields filling the universe. An example is the *Higgs field*, which permeates all space and endows particles their mass. It has a particle associated with it, the Higgs particle, which was discovered at the European Large Hadron Collider in 2012; François Englert and Peter Higgs won the 2013 Nobel Prize in Physics for their theoretical work on the subject. If the Higgs field permeates all empty space, then there can be other fields like it that contribute to the vacuum energy of the universe at early epochs. The amount of vacuum energy present depends on the laws of physics. Guth argued that in the early universe, the weak and strong and electromagnetic forces would have been united in a single superforce, and as a consequence, the vacuum energy at that time might have been very high, unlike today. If this is true, then the cosmological constant was not really a constant as Einstein had supposed but could change with time. Accompanying this high energy density was an equally large negative pressure, ensuring, by the laws of Einstein's theory of special relativity, that the vacuum energy as seen by different observers traveling at different velocities through space would all be the same. Thus, there would be no preferred standard of rest in empty space, as Einstein postulated when he first formulated his theory of special relativity in 1905. This negative pressure would be uniform and exert no hydrodynamic forces. Remember, it takes pressure *differences* to make the wind blow and knock you over. But Einstein's field equations of general relativity from 1915 show that pressure as well as energy density gravitates. The vacuum energy density produces a gravitational attraction, but the *negative* pressure of equal magnitude operates in three directions, producing a gravitational *repulsion* 3 times larger. This vacuum state would manifest as an overall gravitational repulsion, which, according to Einstein's equations, would have started the universe on the accelerated expansion that Guth wanted. It was therefore this gravitational repulsion that produced the initial expansion we call the "Big Bang."

In fact, this trumpet-like solution to Einstein's field equations had already been found by Willem de Sitter in 1917. He solved Einstein's equations for the case of empty space with a cosmological constant and nothing else. We call what he found *de Sitter space*. At late times this space shows exactly the trumpet-shaped spacetime geometry Guth needed for his theory. As measured by clocks carried by particles in this spacetime, the circumference of the universe seems to be doubling in each successive time interval, increasing as 1, 2, 4, 8, 16, 32, 64, 128, 256, 512, 1,024, . . . , resulting in an exponentially accelerated expansion. It's like bad currency inflation, which is why Guth called the model *inflation.*

The separation in space between two particles in this spacetime increases exponentially. Think of a balloon that doubles in size every second as it inflates. Eventually, the distance between two pennies on the balloon will be increasing faster than the speed of light. That does not violate Einstein's theory of special relativity, which just says that someone else's spaceship cannot *pass* you at a speed faster than the speed of light. But general relativity allows the space *between* two particles to stretch so fast that light cannot cover the ever-widening gap between them. De Sitter spacetime explains how particles can be in communication and achieve thermal equilibrium near the bottom of the funnel, the de Sitter waist, and then be spread apart to great distances as the funnel widens with time.

Guth ultimately proposed such a funnel to start the universe, with an initial circumference (the de Sitter "waist") that we estimate to be as small as 10^{-26} centimeters. Guth just needed a tiny volume of high-density vacuum state at the beginning. The repulsive effects of the large negative pressure would cause the spacetime to start expanding, and then to expand faster and faster, with the size of the universe doubling perhaps every 3×10^{-38} seconds (the exact value depending on the details of the model). The universe would become very large in a very short amount of time. As the universe expanded, however, the energy density of the vacuum state would stay the same during this inflationary period. A small region of high energy density would expand to become a large region having the same high energy density.

Curiously, this does not violate *local* energy conservation (i.e., within small neighborhoods), one of the basic principles of modern physics. If I had a small box of high-density, negative-pressure fluid, and I expanded the walls of the box, I would have to do work to pull the walls apart against the negative pressure that was resisting expansion. The work I was doing pulling the walls apart against this negative pressure (or suction) would add energy to the fluid—just enough to keep its energy density at the same high level as the volume of the box expanded. Thus, energy would be conserved locally as required by general relativity. But in the universe, what is pulling on the walls of my box? It is just the negative pressure from the other little similar boxes of spacetime next to it. As long as the pressure is uniform throughout the universe, the expansion itself is doing the work.

In this way the vacuum state is "self-reproducing," growing exponentially large from a tiny beginning. Because of this, Guth noted that the universe "is the ultimate free lunch." Remember, he had suggested that the vacuum state that drove inflation was caused by the superforce unifying the strong, weak, and electromagnetic forces. However, that vacuum state could not last forever and would eventually decay, as the strong, weak, and electromagnetic forces decoupled and became distinct from one another. Meanwhile, as the energy density in the vacuum of empty space dropped, the energy would be converted into a bath of hot elementary particles—exactly the initial conditions Gamow postulated for his hot Big Bang theory.

This is where the inflationary trumpet at the beginning of the universe joins to the bottom of the football-shaped Friedmann Big Bang model. The expansion of the universe then starts to decelerate, as in the football model. The pressure is now just the ordinary thermal pressure of particles, which is positive. Inflation gives a natural explanation of how the initial conditions of the Friedmann Big Bang model were produced. The repulsive gravitational effects of the initial vacuum state (through its negative pressure) started the Big Bang! The Big Bang did not have to start with a singularity but, instead, could begin with a small, high-density vacuum region. Inflation simultaneously explains why the universe was so large, and why it was so uniform. Any wrinkles would be flattened out as the universe stretched to enormous size. It could also explain the small fluctuations of one part in 100,000 that we observe. The universe was doubling in size every 3×10^{-38} seconds in the beginning. On these short timescales, Heisenberg's uncertainty principle of quantum physics ensures random fluctuations in the energy of any field. During inflation these tiny regions will expand enormously to seed the giant cosmic structures we see today. In fact, the spongelike pattern of filaments seen in galaxy clustering (see figure 5) today—the *cosmic web*—as well as the pattern of hot and cold spots in the CMB indicate that the initial conditions were random in precisely the way expected from the quantum fluctuations predicted by inflation

Inflation had one problem, however, that Guth recognized. The high-density vacuum state at the beginning would not be expected to decay into ordinary particles all at once. This high-density inflating sea would decay into separate bubbles of low-density vacuum, a phenomenon investigated by Sidney Coleman. It is like boiling water in a pot. The water does not turn to steam all at once. Bubbles of steam form in the water. But this was not a uniform distribution—not the uniform universe we were hoping for. In 1982, I proposed that inflation would produce bubble universes—each bubble would expand to make a separate

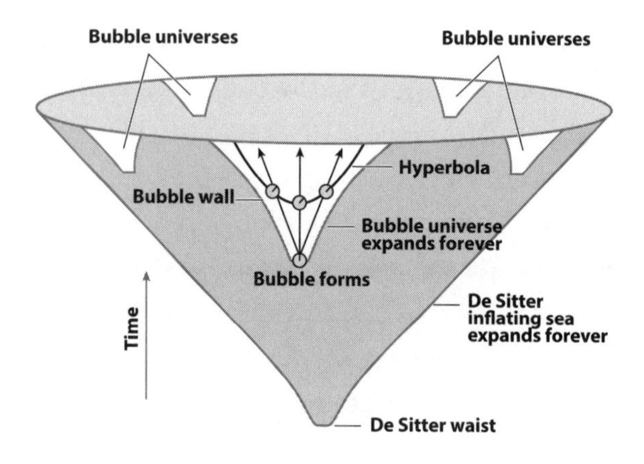

FIGURE 8. Bubble universes forming in an inflating sea—a multiverse.

Credit: Adapted from J. Richard Gott (*Time Travel in Einstein's Universe*, Houghton Mifflin, 2001).

universe like ours (see figure 8). This spacetime diagram shows one dimension of space, the circumference of Guth's inflationary funnel, and the time dimension shown vertically, with the future toward the top. At the bottom, the funnel starts with a small circumference, labeled "De Sitter waist." This represents the circumference (10^{-26} cm) of a tiny, balloon-like, 3-sphere universe. The repulsive gravitational effects of the negative pressure cause the circumference of the funnel to increase more and more rapidly as we go upward toward the future and as the 3-sphere (balloon) inflates.

In my model, our entire observable universe is contained within one of the low-density bubbles. From inside our bubble, we are looking out in space and back in time, so we just see our own bubble universe and the uniform inflating sea before it was created—we don't see the other bubbles. Everything looks uniform to us—solving Guth's nonuniformity problem. The bubble expands at nearly the speed of light. But the inflating sea expands so fast that the bubbles never percolate to fill the entire space. New bubble universes are continually forming, and the inflating sea is expanding between them to provide space for even more new bubble universes to form. This eventually creates an infinite number of bubble universes forming in an ever-expanding inflating sea—what we now call a *multiverse*. These bubble universes would expand forever.

A *surface of constant epoch* in our universe is one for which alarm clocks on individual particles all go off showing the same elapsed time as one another since the bubble formation event. Its shape is hyperbolic, because according to special relativity particles going faster (to the left and right) have clocks that tick more slowly, and therefore, the point at which their alarm clocks go off is delayed. This produces a hyperbolic shape that is infinite in extent as it bends upward inside the expanding bubble wall (as shown in figure 8). This is exactly the geometry that Friedmann's low-density universe of 1924 requires. Eventually, as our bubble expands to infinite volume in the infinite future, an infinite number of galaxies will form inside our bubble. As the inflating sea continues to expand forever outside our bubble as well, an infinite number of infinite bubble universes could be produced from an initially very small, finite, high-density de Sitter space at the de Sitter waist (as indicated in figure 8).

Vacuum energy density—the energy density of empty space—can be viewed as the altitude in a landscape. Different places in the countryside correspond to different values of energy fields (such as the Higgs field) that contribute to the vacuum energy. Different locations, corresponding to dif-

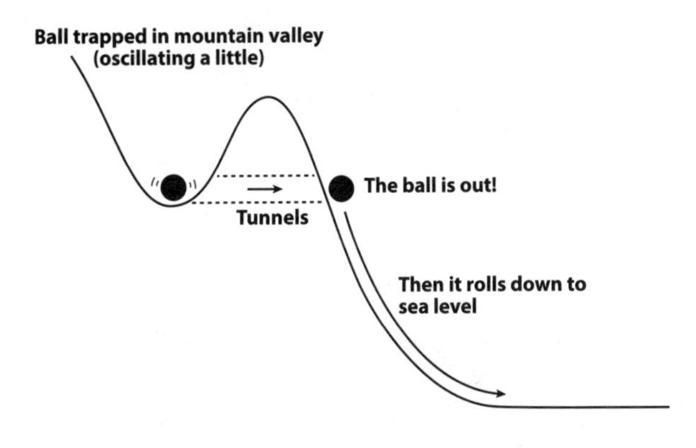

FIGURE 9. Quantum tunneling.

Credit: J. Richard Gott (*Time Travel in Einstein's Universe*, Houghton Mifflin, 2001)

ferent values of the fields, also correspond to different altitudes—different values of vacuum energy density. Today we experience a very low vacuum energy density—we are near sea level. But in the early universe the vacuum energy density was high, like being trapped in a high mountain valley (see figure 9).

A ball trapped in a high mountain valley has a lower energy state it can go to—sea level. But it's trapped if surrounded by mountains on all sides. In Newton's universe, it would have no way to escape and roll down hill, but *quantum tunneling* allows it to exit through the surrounding mountain before descending to sea level.

Bubble universes could have different laws of physics in them if they tunneled, into *different* low-altitude valleys where the values of various fields differed. The laws of physics we see in our particular universe might only be local bylaws, as emphasized by Andrei Linde and Martin Rees.

Quantum tunneling was discovered by George Gamow. On the scale of individual atomic nuclei, it explained the radioactive decay of uranium. Uranium nuclei decay by emitting an alpha particle (a helium nucleus with two protons and two neutrons). That alpha particle is trapped inside the nucleus by the strong nuclear force, which acts like the mountain range surrounding the mountain valley. But the strong nuclear force is only short-range. If the alpha particle could somehow get out of the nucleus, beyond the influence of the strong-force attraction, it could escape. The alpha particle is positively charged and would then be repelled by the positively charged main nucleus. It would roll down the hill, away from the nucleus, and the kinetic energy it picked up would be due to electrostatic repulsion. From the energy an emitted alpha particle has when uranium decays, physicists could calculate how far up the hill it began. It turned out that it was emitted *outside* the uranium nucleus! How did it get out there? Quantum physics tells us that just as light can behave as both a wave and a particle, so too do alpha particles. The wave nature of an alpha particle means that it is not well localized, in a sense captured by Heisenberg's uncertainty principle. Gamow found there was a small probability that the alpha particle could "tunnel" through the mountain that was holding it inside the uranium nucleus and suddenly find itself far outside the nucleus, where it would roll down the hill away from the nucleus owing to electrostatic repulsion. It reminds me of the Zen koan, *how does the duck get out of the bottle* (whose neck is too narrow to allow the duck to escape)? Answer: *The duck is out!* So, the alpha particle quantum tunnels through the mountain, and "the alpha particle is out."

In the bubble universe case, the mountain valley represents the initial inflationary universe, at the de Sitter waist, with its high vacuum energy density. That universe would be happy to stay in a high-density ever-expanding state forever, but after a long time (perhaps many billions of years of inflation), there is a chance it will tunnel through the mountain, where it will roll down to sea level, releasing the energy of the vacuum into kinetic energy and into the creation of elementary particles. This tunneling represents the instantaneous formation of a small bubble with a vacuum energy density inside that is slightly less than the vacuum energy density outside. The negative pressure outside the bubble is stronger than the negative pressure inside the bubble, and the difference pulls the bubble wall outward. It expands faster and faster, eventually approaching the speed of light. Meanwhile, inside the bubble, the vacuum energy density is slowly rolling down the hill toward sea level. Inflation continues for a while inside the bubble while this happens. When it has rolled to sea level and deposited its vacuum energy in the form of particles, inflation stops, and the Friedmann phase begins. It was this type of scenario that Andrei Linde and also Andreas Albrecht and Paul Steinhardt independently published shortly after my paper came out. Outside the bubble, the vacuum state remains up in the mountain valley and an endless inflating sea continues its rapid accelerated expansion (see figure 8 again).

I had discussed the geometry and general relativity involved in the formation of bubble universes by quantum tunneling to make what today we call a "multiverse." I required that inflation continue inside the bubble universe for a while to create the large universe we live in. In the Linde and the Albrecht and Steinhardt models, this occurred naturally as the vacuum energy density in the bubble took some time to roll slowly down the hill toward sea level. Later in 1982, Stephen Hawking published a paper adopting the bubble universe idea and showing that initial quantum fluctuations would be expanded by inflation to appear on cosmological scales, having just the form needed to successfully seed the formation of galaxies and clusters of galaxies. The structure subsequently observed in the distribution of galaxies and in the fluctuations in the CMB is in beautiful accord with predictions from inflation.

Although it is possible for a neighboring bubble universe to collide with ours in the far future (perhaps $10^{1,800}$ years from now, creating a sudden hot spot in the sky whose radiation would probably kill any life around at the time), most of these other universes in the multiverse are forever hidden from our view because they are so far away that the light from them can never cross the ever-inflating region that separates us. It seems pretty clear that once it gets started, inflation is hard to stop. It will continue expanding forever, creating a multiverse with an infinite number of universes, of which ours is one.

Even though we can't see these other universes in the multiverse, we have reason to think they exist, because they seem to be an inevitable prediction of the theory of inflation. And inflation explains a wealth of observational data.

Inflation got a great boost from the measurements of fluctuations in the cosmic microwave background by the WMAP and Planck satellites.

They showed the universe has approximately zero curvature. In a positive curvature universe, like Friedmann's 3-sphere universe of 1922, we would see fewer hot and cold spots in the microwave background map. If it were negatively curved, as in the Friedmann 1924 low-density hyperbolic universe, there would be more spots.

This means that we don't really know the sign of the curvature. The curvature of the universe is just so close to zero that we cannot measure it. Our current data show that the visible universe is flat to an accuracy of 0.2%. In the same way, a basketball court looks flat even though we know it follows the curvature of Earth. It's just that the radius of Earth is very much larger than the basketball court, ensuring that the curvature in the basketball court is not noticeable. Early people thought Earth was flat, because the tiny part of Earth they could see was approximately flat. All we really know is that the radius of curvature of the universe is much larger than the 13.8-billion-light-year radius out to which we can see—out to the CMB. A positively curved balloon-shaped universe would look flat to us if its radius were much larger than 13.8 billion light-years, as would a sufficiently expanded negatively curved hyperbolic bowl-shaped universe as required in Friedmann's 1924 low-density universe model, and as should occur if we live in a bubble universe. Guth emphasized that no matter whether the universe was negatively or positively curved, inflation in the simplest models would typically yield enough expansion to make the universe much larger than the part we can inspect. Guth predicted we would find that the universe was approximately flat, and decades later he was shown to be correct. If ours is a bubble universe, the fact that things look approximately flat simply means that it continued to inflate for a "long" time inside the bubble, as the vacuum state rolled down the hill after tunneling. A "long" period of inflation, say 300 doublings in size as seen from within the bubble, could be accomplished in just 10^{-35} seconds, if the doubling time was 3×10^{-38} seconds. That would make the current radius of curvature of the universe 10^{90} times larger than the part we can see, so it would look flat.

We can get other information on our cosmological model by directly measuring the expansion history of the universe by observing the relationship between redshift and distance to far away objects. Astrophysicists use so-called Type Ia supernovae whose luminosity can be accurately calibrated. These data show the expansion of the universe is currently accelerating, and for this discovery, Saul Perlmutter, Brian Schmidt, and Adam Riess shared the 2011 Nobel Prize in Physics. Models including a substantial amount of positive energy density accompanied by negative pressure in the vacuum of empty space—dark energy—can produce the accelerating expansion we observe today. This vacuum energy density is just like that we invoked for inflation, but at a much lower energy density and thus, working on a much longer timescale (12.2 billion years to double the size of the universe today instead of 3×10^{-38} seconds during inflation in the early universe). Data today from the CMB, supernovae, large-scale structure, motions of galaxies, and the nucleosynthesis of light elements in the early universe all converge on a unique inflationary Big Bang model composed of 5% ordinary matter, 27% dark matter, and 68% dark energy.

Sometimes people say that dark energy is a mysterious force causing the current acceleration of the expansion of the universe, or that we know nothing about dark energy. That's not really true. We actually know a lot about dark energy. We know that its energy density is positive, because positive energy density, above and beyond that of ordinary matter and dark matter, is required to make the universe approximately flat, which we observe. We know that the energy density of dark energy is equivalent to about 6×10^{-30} grams per cubic centimeter (using Einstein's famous equation $E = mc^2$). We know that the pressure of dark energy is negative, because, given that the energy density of dark energy must be positive, only a negative pressure for the dark energy could produce the gravitational repulsion required to cause the accelerating expansion of the universe we observe.

The WMAP and Planck satellite results, when combined with supernovae and other data, accurately tracked the expansion history of the universe and, through application of Einstein's equations, establish the *ratio of the pressure to the energy density* in the dark energy today, a key measure that is simply called w_0. Using a fitting formula developed by Zack Slepian and me, the Planck satellite team has found that $w_0 = -1.008 \pm 0.068$. This estimate is consistent, within the observational uncertainties, with the value of $w_0 = -1$ expected from dark energy not changing with time. Einstein should be happy. His cosmological constant term was a brilliant idea after all!

The force that is causing the accelerated expansion of the universe is just *gravity*. And it's repulsive because of the *negative* pressure associated with dark energy. We suspect that what we call dark energy is a form of vacuum energy produced by a field or fields, but we don't know which one or ones. We know that the amount of dark energy is approximately constant with time, but we don't really know whether it is slowly falling (rolling down the hill) or rising (rolling up the hill). This is the focus of current research.

A new independent test for inflation in the early universe has recently been proposed. If inflation causes the universe to double in size approximately once every 3×10^{-38} seconds, then the visible universe in the inflationary epoch would extend out to a distance of only 3×10^{-38} light-seconds or 10^{-27} centimeters. This distance is tiny, and in agreement with Heisenberg's uncertainty principle of quantum physics, this causes fluctuations in the geometry of spacetime (ripples), which, according to Einstein's equations, travel at the speed of light. These propagating fluctuations are gravitational waves. These would leave a characteristic swirling pattern in the polarization of the microwave background radiation, which could in principle be measured. But so far, its detection has proved elusive. The inflationary model the Planck satellite team thinks fits the data best is one by Alexei Starobinsky. Its doubling time of 3×10^{-38} seconds at the end of the inflationary epoch would produce gravitational waves of amplitude safely below the current upper limits on our capacity to measure them. A number of observational efforts, including high-altitude balloons and ground experiments in Antarctica, are underway to lower the observational uncertainties and further test inflationary models. Astrophysicists are anxiously waiting to see whether these observations can open a new window on the early universe.

Inflation has been very successful at explaining the structure of the universe that we see. We don't know for sure how inflation got started, because inflation "forgets" its initial conditions as the universe exponentially expands, thinning out any initial components. But there are some speculations.

As we've seen, inflation can begin with a tiny de Sitter "waist" universe with a circumference of perhaps only 10^{-26} cm, which will then start expanding. But where did *that* come from? Alex Vilenkin has proposed it might simply pop into existence from nothing via quantum tunneling, a process analogous to that occurring with the formation of bubble universes. This time, the ball at rest in the mountain valley would correspond to a universe of zero size. It would then tunnel through the mountain to find itself suddenly outside on the slope. This would correspond to a finite-sized universe—the de Sitter waist. Then as it rolled down the hill, that phase would correspond to the de Sitter funnel (see figure 8). James Hartle and Stephen Hawking have proposed a model with similar geometry.

Quantum tunneling is certainly weird. We are looking for something weird to happen at the beginning of the universe, because what happened then was truly remarkable. Maybe it could have been quantum tunneling. But you don't really start with nothing. You start with a quantum state corresponding to a universe of size zero that knows all about the quantum laws of physics. But how does *nothing* know about the laws of physics? The laws of physics are simply the rules by which stuff behaves. With no stuff, what do the laws of physics mean? So you face this problem trying to make a universe out of nothing.

Meanwhile, Linde noted that an inflating universe can give birth to another inflating universe via a quantum fluctuation. A de Sitter inflating trumpet horn could give birth to another inflating trumpet horn, which would sprout from it, the way branches grow off a tree. A branch will then inflate and grow to be as large as the trunk, and sprout branches of its own. Branches will continue forming branches, making an infinite fractal tree of universes, all from one original trunk. Each individual branch is a funnel that could form bubble universes (as in figure 8). We would exist in a bubble universe on one of the branches, but still, you might ask: where did the trunk come from?

Li-Xin Li and I tried to answer this question. We proposed that one of the branches curled back in time and grew up to be the trunk. When a branch curls back in time and grows up to become the trunk, it creates a little time loop in the beginning that looks like the loop in the number "6" (see figure 10).

The four inflationary funnels labeled 1, 2, 3, 4 are four of Linde's branch inflationary universes, each of which will form bubble universes as in figure 8. Universe 2 gives birth to universes 1 and 3. Universe 4 is its grandchild universe. Because of the time loop at the bottom, universe 2 is its

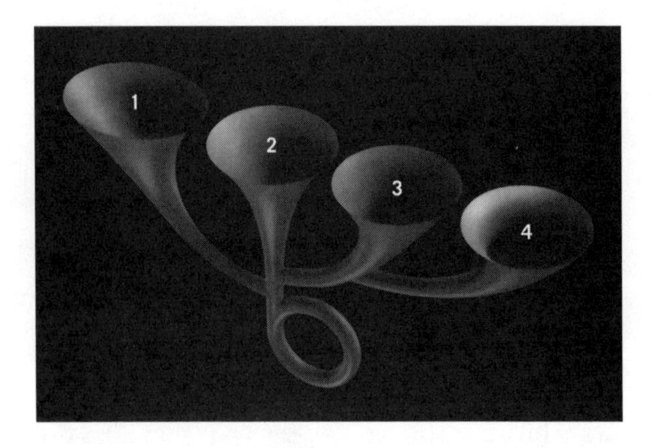

FIGURE 10. Gott-Li self-creating multiverse. The loop at the bottom represents a time machine; the universe gives birth to itself.

Photo credit: J. Richard Gott, Robert J. Vanderbei (*Sizing Up the Universe*, National Geographic, 2011).

own mother. Such time loops are possible in general relativity: for example, wormhole solutions found by Kip Thorne have them, as do also my solutions involving cosmic strings. So this universe has a little time machine at the beginning—every event has events that precede it. If you find yourself within this short time loop, which may range from 5×10^{-44} seconds to 10^{-37} seconds long, there are always events counterclockwise from you that are before you and give rise to you in the usual causal way. This multiverse is finite to the past but has no earliest event.

One of the wonderful things about inflation is that a small piece of inflating vacuum state expands to create a large volume, each little piece of which looks exactly like the piece you started with. If one of those little pieces *is* the piece you started with, then you have a time loop. In the Gott-Li theory, the universe isn't created from nothing, but rather it is created from something, a little piece of itself. The universe can be its own mother.

Today, the theory of inflation is in very good shape. It explains the fluctuations we see in the CMB in detail. If you doubt that inflation occurred, remember that we see a low-grade version of inflation going on today. The universe's expansion is accelerating, caused most likely by a dark energy vacuum state with a tiny density of 6.9×10^{-30} grams per cubic centimeter. Inflation just relies on a large amount of dark energy in the early universe, and seems likely to produce a multiverse. What a spectacular prospect. Research on all this continues.

CHAPTER 8

OUR FUTURE IN THE UNIVERSE

J. Richard Gott

The universe today is 13.8 billion years old. By the time its age is 19 billion years, the Andromeda galaxy will have crashed into the Milky Way. By 22 billion years after the Big Bang, the Sun will have finished its main-sequence lifetime and will have already become a white dwarf.

As we have discussed, observations indicate that the universe is filled with dark energy characterized by a pressure equal in magnitude to its energy density but *negative*—dynamically equivalent to Einstein's cosmological constant. As the matter of the universe thins out because of the expansion, the dark energy nonetheless remains at the same density, forcing the universe to become ever more dominated by dark energy in the far future. This means a new epoch of low-grade inflation is starting up again. The Belgian physicist-priest George Lemaître was the first to consider such a model for the future of our universe. As time goes on, two galaxies that can communicate today will flee from each other on the fabric of spacetime faster and faster. Eventually the space between the two galaxies will expand so fast that light cannot cross the ever-increasing distance between them. Event horizons form. A distant galaxy will look to us just like it is falling into a black hole, forever concealing from us all events occurring at late times in the distant galaxy.

At about 850 billion years, by which point dark energy completely dominates the universe's expansion, the universe will cool to a constant temperature, owing to a process described by Gibbons and Hawking. Hawking had earlier showed that event horizons create Hawking radiation. Black holes have event horizons that don't allow you to see inside. This causes Hawking radiation to be emitted from their surface, which ultimately causes them to evaporate. But if the universe continues to expand in an accelerated fashion, event horizons will also form because there are distant events we will never see. Gibbons and Hawking calculated that in such an inflating universe at late times, any observers present would see the resulting thermal radiation: *Gibbons and Hawking radiation*. This thermal radiation, seen in the future of our universe, would have a characteristic wavelength of about 22 billion light-years. The CMB radiation continues to increase in wavelength as the universe expands exponentially, doubling its size every 12.2 billion years. After 850 billion years, the CMB thermal radiation will have a characteristic wavelength longer than 22 billion light-years and will become negligible compared with the Gibbons and Hawking radiation produced by the event horizons. At that point we should see the temperature of the universe stop falling and become constant at a Gibbons and Hawking temperature of about 7×10^{-31} K. That's very cold, but still above absolute 0 K.

This is ultimately bad for intelligent life, if any exists 850 billion years from now. As other galaxies flee from us and disappear behind an event horizon, we would be left alone with finite energy resources, forever stuck in a low-temperature bath. Living creatures need energy to live, but eventually we would run out of energy and come into equilibrium with this thermal bath. Intelligent life can in principle last a very long time by huddling around evaporating black holes, warming themselves and getting energy from their own Hawking radiation. However, even the most massive black holes will eventually evaporate (about 10^{100} years from now), and intelligent life will die out. So, at 850 billion years, when the Gibbons and Hawking radiation begins to dominate, intelligent life will be given fair warning of distant future trouble.

These ideas are actually testable. Gibbons and Hawking radiation is also produced in the early inflationary phase of the universe. This includes both electromagnetic radiation (light) and gravitational radiation. If such gravitational radiation from the early universe is eventually detected via the imprint left on the polarization of the CMB, it would constitute an important experimental verification of the Hawking radiation mechanism. These gravitational waves are not made by moving bodies, like the colliding black holes recently detected by LIGO; these would be produced by something different, the Hawking mechanism—a quantum process. So this would be something new and exciting.

Here's more trouble for life in the distant future. At 10^{14} years, the stars fade as the last low-mass stars run out of hydrogen fuel and die. The universe becomes dark. Only stellar corpses remain—white dwarfs, neutron stars, and black holes. Some planets may still circle them. But by 10^{17} years, enough close encounters of stars will have occurred to rip the planets from their orbits and fling them into interstellar space.

At 10^{21} years, galactic-mass black holes form. Two-body gravitational interactions slingshot some stars out of galaxies, while the rest fall into the central black hole. Gravitational radiation causes stars close to the black hole to spiral in.

By 10^{64} years (if it hasn't happened already), according to Hawking, protons should decay through a rare process of temporarily falling inside a microscopic black hole (via the uncertainty principle), and then having the black hole decay quickly by Hawking radiation. To conserve charge, a positron, which is much lighter than the proton, can be emitted as one of the decay products of the black hole into which the proton has disappeared. As protons decay, we are left with electrons and positrons as the most massive remaining particles. Protons may decay even earlier than this; perhaps on a timescale of 10^{34} years, but regardless of what happens then, they would have all decayed in any case by 10^{64} years. At 10^{100} years, galactic-mass black holes evaporate via Hawking radiation and are gone.

What happens after that?

The standard picture physicists have is that dark energy, which today is causing an exponential expansion of the universe, represents a vacuum state with a constant positive energy density, and a negative pressure. As we have discussed in the previous chapter, we can liken our current situation to living in a valley that sits slightly above sea level—our altitude indicating the amount of dark energy present in the vacuum. We have rolled to the bottom of this valley and are just sitting there. The amount of energy in the vacuum—the dark energy—is not changing. This will keep the universe doubling in size every 12.2 billion years for a very long time.

Wait long enough, and our vacuum state, which causes dark energy, could quantum tunnel through the valley walls into a lower energy state (of lower-altitude terrain beyond our valley). This would cause a bubble of lower-density vacuum state to form somewhere in our visible universe. The negative pressure outside the bubble would be more negative than the pressure inside, which would pull the bubble wall outward. After a short time, the bubble wall would be traveling outward at nearly the speed of light and would expand forever. The laws of physics would be different inside that bubble, and you would be killed when that bubble's wall hit you.

We could possibly see bubbles of lower-density vacuum forming in as "little" as 10^{138} years owing to a known instability in the Higgs field vacuum energy. But many physicists think the Higgs vacuum will be stabilized by higher-energy effects. In that case, according to speculative calculations by Andrei Linde, bubbles of lower-density vacuum should start forming only after 10^(10^34) years. These bubbles would form, and just like the bubble universes in figure 8, they would never percolate to fill the entire space. The ever-expanding vacuum state would continue to double in size every 12.2 billion years and have a volume that would increase endlessly—an ever-inflating sea punctuated by forming bubbles. At late times, our universe resembles eternally fizzing champagne.

Alternatively, we don't live in a valley at all, but on a slope, and we will slowly roll down to sea level. This is called *slow-roll dark energy*. As Bharat Ratra, Jim Peebles, Zack Slepian, and I, and many others, have explored, this would cause the amount of dark energy to slowly dissipate over billions of years, rolling down ultimately to a vacuum state of zero energy density. Just such a rolling down occurred once before with inflation, where a very high-density dark energy state rolled down to the low-energy vacuum that we see today. That could occur again, allowing us to ultimately roll down to sea level—a vacuum energy of zero. In that case the Gibbons and Hawking radiation would not occur, and in principle intelligent life could last a much longer time. These scenarios can be investigated by carefully measuring the expansion history of the universe up to now. This allows us, using Einstein's equations, to measure the ratio of the pressure to energy in the dark energy today, the quantity we defined as w_0 in chapter 7. If w_0 turns out to be exactly –1, dynamically equivalent to Einstein's cosmological constant, which favors the "trapped in a valley" scenario, the dark energy will remain at its present value, and the universe will keep doubling in size every 12.2 billion years forever, with quantum tunneling ultimately producing bubbles of lower dark energy density over long timescales. If w_0 is less negative than –1 (i.e., between –1 and 0), however, we could roll slowly down to sea level, and the accelerated expansion should eventually give way to an approximately linear expansion rate. In that case, the universe would continue to expand forever but at a linear rate: going like 1, 2, 3, 4, 5, 6, . . . with time.

A radical proposal by Robert Caldwell, Mark Kamionkowski, and Nevin Weinberg is that w_0 could be more negative than –1. This model is called *phantom energy*. It would produce a vacuum energy that *increased* with time as the universe expanded, leading to a runaway expansion and creating a singularity—a state of infinite curvature and infinite tidal forces—in the future (a Big Rip) that would tear galaxies, stars, and planets apart in perhaps as few as a trillion years from now. Although that remains a possibility, it seems less likely than the other two scenarios. But many physicists take "phantom energy" quite seriously.

Which of these three scenarios is relevant? It depends on the value of w_0, which is the ratio of pressure to energy density in dark energy today. As we have mentioned, the best estimate of w_0 is, within the uncertainties, consistent with the simple value of –1, which corresponds to the model where the universe is sitting at the bottom of a valley. This result strongly supports the general idea that dark energy represents a vacuum state with positive energy and negative pressure, but these observations are not yet able to truly distinguish between models in which we are sitting still at the bottom of a valley from those in which we are slowly rolling down (or up) a hill. In the latter cases, w_0 would be close to but not exactly equal to –1, slightly above (or below) it. If future, high-precision measurements of w_0 show that it is unambiguously different from –1, we could learn whether the slow-roll dark energy or phantom energy picture was favored. But, if, as measurements continue to improve and the uncertainties continue to go down, and we continue to be consistent with $w_0 = -1$, we may well pronounce the "sitting at the bottom of the valley" model triumphant. There are a number of experimental programs either underway or proposed that can potentially lower the uncertainties in w_0 by perhaps a factor of 10, which would further illuminate the ultimate fate of our universe.

Now you have our best predictions of what the universe will do in the future. But what about *our* future in the universe? What's likely to happen to us? How is our species *Homo sapiens* likely to fare in the far future?

Note first that we are living in a very habitable epoch. The universe is old enough and has cooled off enough for carbon and other elements essential to life to have formed. And the stars are shining nicely today, providing warmth and energy. Just the kind of epoch in which we might expect to find intelligent observers. After the stars have faded, it will be more difficult for intelligent life to arise or thrive. The *Weak Anthropic Principle*, proposed by Robert Dicke, and later given its name and precise formulation by Brandon Carter, says that intelligent observers should, of course, expect to find themselves at habitable locations—in a habitable epoch in the universe. Why? Because logically, they wouldn't be alive to be asking the question in an *uninhabitable* epoch.

As the only intelligent observers we have encountered in the universe so far, we would like to know how long our tenure as a species is likely to last. How might we think about this question?

In 1969, I visited the Berlin Wall, which separated the two sectors of the city belonging to East and West Germany. People at that time wondered how long the Berlin Wall would last. Some people thought it was a temporary aberration and would be gone quickly. But others thought the wall would remain a permanent feature of modern Europe.

To estimate the wall's future longevity, I decided to apply the Copernican Principle—the idea that our location is not likely to be special—a concept that we have encountered earlier. I thought: I'm not special. My visit is not special. I am just coming to Europe after college. I'm visiting the Berlin Wall just because I'm in Berlin and the wall happens to be there. I could have seen it at any point in its history. But if my visit is not special, my visit should be located at some random point between the wall's beginning and its end. The end officially comes either when the wall ends or when there is no one left alive to see it, whichever comes first. There should be a 50% chance, then, that I am located somewhere in the middle half of its existence—in the middle two quarters. If I were visiting at the beginning of that middle 50%, then I would have been 1/4 of the way through its history, with 3/4 still in the future. In this case, the wall's future longevity would be 3 times its past longevity. In contrast, if I were at the end of that middle 50%, 3/4 of its history would be past and 1/4 would remain in the future, making its future 1/3 as long as its past. There is a 50% chance that I would be between these two limits and that the future longevity of the wall would be between 1/3 and 3 times as long as its past.

At the time of my visit, the wall was 8 years old. While standing at the wall, I predicted to a friend, Chuck Allen, that the future longevity of the wall would be between 2 2/3 years and 24 years.

Twenty years later I was watching television and called up my friend saying: "Chuck, you remember that prediction I made about the Berlin Wall? Well turn on your television, because NBC news anchor Tom Brokaw is at the wall now and it's coming down today!" Chuck did remember the prediction. The Berlin Wall had come down

20 years later, within the range of 2 2/3 years to 24 years that I had predicted. My visit was in the middle of the Cold War, so an atomic bomb could have taken it (and me) out in the next millisecond. In contrast, some famous walls, such as the Great Wall of China, have lasted for thousands of years. My predicted range was rather narrow but still gave me the right answer.

Scientists generally prefer to make predictions that have a 95% chance of being correct. That is the usual 95% confidence level used in scientific papers. How does this change the argument? When applying the Copernican Principle, keep in mind that if your location in time is not special, there is a 95% chance that you are somewhere in the middle 95% of the period of observability of whatever you are observing—that is, neither in the first 2.5% nor in the last 2.5%.

Expressed as a fraction, 2.5% is 1/40. If your observation falls at the beginning of that middle 95%—only 2.5% from the start—then 1/40 of the history of what you are observing is past, and 39/40 of it remains in the future. In this case, the future is 39 times as long as the past. If you are only 2.5% from the end, then 39/40 of it is past and 1/40 remains. Thus:

The future longevity of whatever you are observing will be between 1/39 of its past longevity and 39 times its past longevity (with 95% certainty).

I decided I would like to apply this to something important, to the future of the human species, *Homo sapiens*. In our current form, we are about 200,000 years old. That goes back to just before Mitochondrial Eve, in Africa, from whom we are all descended. The formula would say with 95% confidence that, if our location in the timeline of the history of our species is not special, the future longevity of our species *Homo sapiens* should be at least 5,100 more years (that's 200,000/39) but less than 7.8 million more years (that's 200,000 × 39). We do not have actuarial data on other intelligent species (those able to ask such questions), so arguably this is the best we can do. The range of predicted future longevity is as large as this, because one wants to be 95% certain of being correct. Yet many experts who offer their own estimates make predictions outside this range. Some apocalyptic predictions say we are likely to be extinct in less than 100 years. But if that were true, we would be very unlucky to be located at the very end of human history. Some optimists think we will go on to colonize the galaxy and last for trillions of years. But if that were true, we would be very lucky to be located at the very beginning of human history. So even with its broad range, the Copernican-based formula is still highly informative, limiting the possibilities to a tighter range than those considered by many others.

Interestingly, the total longevity of our species predicted by the Copernican formula agrees remarkably well with the actual longevities of other species on Earth. *Homo erectus*, our parent species, lasted 1.6 million years, and the Neanderthals lasted only about 300,000 years. Mammal species have an average longevity of 2 million years, and other classes of species have average species longevities of between 1 and 10 million years. Even the fearsome *Tyrannosaurus rex* went extinct after existing for only 2.5 million years. It was knocked out by an asteroid strike about 65 million years ago.

If we resorted only to actuarial data on other mammal species to forecast our future longevity, we would find a future longevity of between 50,600 years and 7.4 million years (with 95% confidence). These limits are within the those derived from the Copernican Principle, based only on our past longevity as an intelligent species. As long as we stay on Earth, we are subject to the same dangers that have caused other species to go extinct, and the fact that we have been around only 200,000 years should make us worry that our intelligence will not necessarily improve our fate relative to other species.

Now you might think, sure, *Homo sapiens* will go extinct, but that's okay, because we will give birth to an even more intelligent species in the future to replace us. Darwin noted, however, that most species leave no descendant species at all. They die off without progeny. In this regard, note that all other species in our hominid family—including the Neanderthals, *Homo heidelbergensis*, *Homo erectus*, *Homo habilis*, and *Australopithecus*—have gone extinct. We are the only hominid species left. The rodent family, by comparison, has 1,600 different species alive today and has many chances for survival. Stephen Jay Gould made a similar point, calling us "just one bauble" on the Christmas tree of evolution. We shouldn't count on leaving descendants that are intelligent machines either. If most intelligent creatures in the history of the universe are intelligent machines, *and* you are not special, then you must ask yourself: "Why am I not an intelligent machine?"

Some people, like Oxford philosopher Nick Bostrom, think we may be intelligent simulated creatures on some advanced future human's laptop. If computer technology continues to advance exponentially, one day future humans may make hyper-realistic simulations that would fool the simulated humans into believing they were real. And, if such simulated humans far outnumbered real humans, you might expect to be one of them, so the argument goes. However, William Poundstone, in his 2019 book *Doomsday Calculations*, has explained why this argument isn't as strong as it first appears—it doesn't really work. In such an optimistic scenario, you might expect to be a simulated human. But the eons of *future* real humans living in the advanced technological epoch of hyper-realistic simulations would be expected to vastly outnumber real humans living before that. Most simulated humans would therefore expect to find themselves (like the majority of real humans they were simulating) experiencing "life" in an epoch when such hyper-realistic simulations were being conducted. You do not, a fact that would make you special (i.e., improbably rare). Therefore, the simulation hypothesis is improbable on Copernican grounds—in such scenarios you would be unlikely (simulated or not) to find yourself in a pre-hyper-realistic simulation epoch because most people (simulated or not) do not find themselves living there.

For example, plays are simulations. If you are an actor in a play, you are likely to be performing in a play set in an epoch when plays already exist—that is, after the ancient Greeks. Likewise, most movies are set in epochs in which movies exist. If the simulation hypothesis were true, you would be special—one of the relatively few intelligent creatures who think they live in an epoch before hyper-realistic simulations were possible. David Kipping has pointed out that the likelihood of you being a simulated human must be even further reduced because we have no idea whether such hyper-realistic simulations are even possible or practical.

Of course, there are certain situations in which my Copernican formula shouldn't be used. Don't use it at a wedding one minute after the vows have been said, to forecast that the marriage will be over in only 39 minutes. You have been invited to the wedding at a special time to witness its beginning, so you are not a random observer. But most of the time you *can* use the Copernican formula. Since my formula was introduced in 1993, it has been tested many times and successfully predicted the future longevities of everything from Broadway plays and musicals, to governments, to reigns of world leaders. Another exception: do not use it to estimate the future longevity of the universe. You may live in a special, habitable, location, because you are an intelligent observer—the essence of what is called the *An-*

thropic Principle. Intelligent observers were not present in the hot early universe and may die out when main sequence stars burn out. But *among* intelligent observers your location in spacetime should not be special. In general, the Copernican formula works because out of all the places for intelligent observers to be, there are by definition only a few special places, and many nonspecial places. You are simply likely to live in one of those many nonspecial places. Also, your current observation is not likely to be in a special location relative to the full array of observations made by intelligent observers. Brandon Carter, of Anthropic Principle fame, and John Leslie and Holgar Nielson have independently come to similar conclusions.

I was able to use the Copernican idea, and some fancy algebra, to set an upper limit on the mean longevity of radio-transmitting civilizations: 12,000 years (with 95% confidence). If the mean longevity were longer than that, my 1993 *Nature* paper announcing my Copernican predictions would appear either unusually early in the history of our radio-transmitting civilization, or unusually early in the timelines of all radio-transmitting civilizations added together and ranked by longevity. This yields a Copernican estimate that you can plug into the Drake equation: $L_C < 12{,}000$ years (with 95% confidence). We used this estimate in chapter 4.

If you think intelligent species typically colonize their galaxy, then ask yourself—why am I not a space colonist? In 1950, Enrico Fermi asked a famous question about extraterrestrials: *Where are they?* Why haven't they already colonized Earth, long ago? The Copernican Principle offers an answer to Fermi's question: A significant fraction of all intelligent observers must still be sitting on their home planets (otherwise you would be special). Colonization must not occur that often. Importantly, this means that we are allowed to apply the Drake equation in the first place, which estimates the number of intelligent civilizations that arise independently on their home planets.

What is the chance that human beings will ultimately colonize the entire Milky Way—1.8 billion habitable planets, the number we found in chapter 4? The Copernican argument says that if you are not special, there is only one chance in 1.8 billion that you would find yourself in the first 1/1.8 billion of all planets ever inhabited by humans and therefore only one chance in 1.8 billion that we would ultimately go on to colonize 1.8 billion planets, given that you are on the first one. Nevertheless, our colonizing a few more planets in the future, starting with Mars, would not be that improbable and could give us more chances to survive. We should be doing that quickly, while we still have a space program.

The goal of the human spaceflight program should be to improve the survival prospects of our species by colonizing space.

We are six decades into the human spaceflight program. The Copernican Principle suggests that funding for the human spaceflight program has a 50% chance of lasting for at least another 60 years—long enough to establish a Mars colony. Asking for such a Mars colony is not unreasonable. My Copernican argument suggests we should do it in a hurry before money for the human space program runs out. In a 2010 interview with bigthink.com, Stephen Hawking put it this way: "The human race shouldn't have all its eggs in one basket, or on one planet. Let's hope we can avoid dropping the basket until we have spread the load."

Elon Musk, head of SpaceX, is interested in privately funded efforts to colonize Mars.

Colonizing Mars would give our species two chances instead of one and might as much as double our long-term survival prospects. It would be a life insurance policy against any catastrophes that might overcome us on Earth, from warfare, to climatological disasters, to asteroid strikes, to surprise lethal pandemics. It might also double our chances of ever getting to another star system. Colonies can spawn other colonies. The first words spoken on the Moon were in English. Not because England sent astronauts to the Moon, but because it founded a colony on North America that did.

If we look around, we can see the universe showing us what we should be doing. We live on a tiny speck in a vast universe. The universe tells us: spread out and increase your habitat to improve your survival prospects. We live on a planet littered with the bones of extinct species, and the age of our species is tiny relative to that of the universe as a whole. We should spread out before we die out. We have a space program only a little over a half a century old that is capable of sending us to other planets. We should make the wisest possible use of it before it is gone. Will we venture out, or turn our backs on the universe? The fact that we are having this conversation on Earth is a warning that there is a significant chance that we will end up trapped on Earth.

Our intelligence gives us great potential, the potential to colonize the galaxy and become a supercivilization, but most intelligent species must not have been able to achieve this—or you would be special to find yourself still a member of a one-planet species. The energy sources we control are far less powerful than our own Sun. We are weak, and we have not been around for very long. But we are intelligent creatures, and we have learned a lot about the universe and the laws that govern it—how long ago it started, how its galaxies and stars and planets formed.

A stunning accomplishment whose story we have told here.

ACKNOWLEDGMENTS

This book came to fruition thanks to the hard work of many people. To start with, we thank our fellow faculty at Princeton University, from whom we have learned so much over the years, and who have given us such a productive and congenial atmosphere in which to work. We especially thank Professor Neta Bahcall.

We thank our students, including Cullen Blake, Wes Colley, Julie Comerford, Daniel Grin, Yeong-Shang Loh, Justin Schafer, Joshua Schroeder, Zack Slepian, Iskra Strateva, and Michael Vogeley. We thank Ramin Ashraf, Sorat Tungkasiri, Paula Brett, Sofia Kirhakos Strauss (Michael's wife), and Kathy Gryzeski for help along the way, as well as Lucy Pollard-Gott (Rich's wife), who copy-edited the entire book. We thank Robert J. Vanderbei and Li-Xin Li for help with graphics. We also thank Adam Burrows, Chris Chyba,

Matias Zaldarriaga, Robert J. Vanderbei, and Don Page for helpful conversations.

At Princeton University Press, we thank our production editor, Mark Bellis, our copyeditor, Kathleen Kageff, and, for her extraordinary faith and vision, our editor, Ingrid Gnerlich.

We thank our wives, children and grandchildren for their love.

Michael A. Strauss
Neil deGrasse Tyson
J. Richard Gott

INDEX

Index note: Page numbers in *italic* indicate illustrations.